A Significant Gap

David Irvine

Printed in the United States of America

First Printing, 2017

ISBN 978-0-9984982-1-8

QCG Press
Baltimore, Maryland

DEDICATED TO THE REAL MOM AND THE REAL SAM

Contents

Prologue

ONE SATURDAY AFTERNOON in the early summer of 1956, fresh out of the second grade, I sat in the Upper Alton Theatre on Washington Avenue, eating popcorn with a close friend and settling in for the weekly movie. *Forbidden Planet* was just beginning.
A spaceship moved across a field of stars. A deep, resonant voice announced,

> In the final decade of the 21st century, men and women in rocket ships landed on the moon. By 2200 AD they had reached the other planets of our solar system. Almost at once there followed the discovery of hyperdrive through which the speed of light was first attained and later greatly surpassed. And so at last mankind began the conquest and colonization of deep space. United Planets Cruiser C-57D is now more than a year out from Earth base on a special mission to the planetary system of the great main sequence star, Altair.

C-57D landed on the apparently barren fourth planet of the Altair system, which was actually the ancient home of an advanced alien race – the Krell. The depiction of that landing on an alien desert world was a masterpiece of cinematography; it would compare favorably with contemporary special effects. Years later, as an aerospace engineering student at Purdue University, I was walking through the Student Union one evening when I chanced upon that same opening scene – *Forbidden Planet* was one of the "What? Me Study?" movies of finals week. Standing outside the door, I saw C-57D in deep space again; I heard the opening lines again; I saw the ship land again. Postponing my review for an impending orbital mechanics test, I sat down and watched the entire movie.

A fascination with deep space had led me into aerospace engineering. Once in college, I discovered another fascination: deep time. One spring day, buried in a back corner of the library, I came upon *The Immense Journey*, by Loren Eisley – a poet as well as an anthropologist. The first essay was entitled "The Slit":

> Some lands are flat and grass-covered, and smile so evenly up at the sun that they seem forever youthful, untouched by time. Some are torn, ravaged and convulsed like the features of profane old age. Rocks are wrenched up and exposed to view; black pits receive the sun but give back no light.
>
> It was to such a land I rode, but I rode to it across a sunlit, timeless prairie over which nothing passed but antelope or a wandering bird. On the verge where that prairie halted before a great wall of naked sandstone and clay, I came upon the Slit. A narrow crack worn by some descending torrent had begun secretly,

far back in the prairie grass, and worked itself deeper and deeper into the fine sandstone that led by devious channels into the broken waste beyond. I rode back along the crack to a spot where I could descend into it, dismounted, and left my horse to graze.

The crack was only about body-width and, as I worked my way downward, the light turned dark and green from the overhanging grass. Above me the sky became a narrow slit of distant blue. In that sky, several stars emerged from the unseasonable twilight. The sandstone was cool to my hands on either side. The Slit was a little sinister – like an open grave, assuming the dead were enabled to take one last look – for over me the sky seemed already as far off as some future century I would never see.

I ignored the sky, then, and began to concentrate on the sandstone walls that had led me into this place. It was tight and tricky work, but that cut was a perfect cross section through perhaps ten million years of time. I hoped to find at least a bone, but I was not quite prepared for the sight I finally came upon. Staring straight out at me, as I slid farther and deeper into the green twilight, was a skull embedded in the solid sandstone. I had come at just the proper moment when it was fully to be seen, the white bone gleaming there in a kind of ashen splendor, water worn, and about to be ground away in the next long torrent.

> It was not, of course, human. I was deep, deep below our time, in a remote age near the beginning of the reign of mammals. I squatted on my heels in the narrow ravine, and we stared a little blankly at each other, the skull and I. It was the face of a creature who had spent his days following his nose, who was led by instinct rather than memory, and whose power of choice was very small. Though not human, nor a direct human ancestor, there was yet about him, even in the bone, some trace of that low, snuffling world out of which our forebearers had so recently emerged. The skull lay tilted in such a manner that it stared, sightless, up at me as though I, too, were already caught a few feet above him in the strata and, in my turn, were staring upward at that strip of sky which the ages were carrying farther away from me beneath the tumbling debris of falling mountains. The creature had never lived to see humans, and I, what was it I was never going to live to see?[1]

Deep time touched me in much the same way that deep space always had; it awakened an indistinct, indefinite but nonetheless deep and powerful yearning. Fortunately for me, Purdue's engineering college wisely required its students to take non-technical courses. This gave me a chance to feed my yearning in a way that stolen hours in the library could not. I was especially drawn to philosophy. I didn't learn much about yearning, but I learned a lot about knowing. Epistemology – the study of knowledge – is one of the main branches of modern philosophy: How do we get knowledge?

How do we validate it? Does it truly tell us about the real, live, lived-in world?

Back in the fifties and sixties, science and philosophy got along better than they do today. What was known as logical positivism was still ascendant, though about to go into decline. The hallmark of positivism was the claim that only empirically verifiable statements were meaningful. Science relied on empirical verification. Therefore, the statements of science were meaningful. A strong empirical foundation was (and still is) a real strength of modern science. Positivism was an appealing philosophy for an aerospace engineer; I was an instant convert. But dependence on empirical verification was (and still is) a real weakness too. Its inevitable corollary is a deeply-rooted tentativeness; the verified data with which you confirm hypotheses today are always subject to newer, more extensive (verified) data that disconfirm them tomorrow. The claims of science are tethered to the empirical experience accessible to scientists.

Positivism has gone out of fashion (though Stephen Hawking still calls himself a positivist), but philosophers are still a skeptical lot, and the passing of positivism by no means diminished their skeptical critique of knowledge. More recent philosophers have preferred tethering knowledge to social usefulness; a widespread formulation has it that knowledge is socially constructed. But whether the reins are held by society or by science, the instinct to rein in the pretensions of knowledge is still alive and well. To wit: Before I can claim to know something, both my claim and that *something* must satisfy definite standards.

On one hand, I have no problem with this. The close tie of aerospace engineering to scientific knowledge is attested by its curious nickname: Rocket Science. Rockets depend upon rigorous science. Aerospace engineers firmly believe in right answers and wrong answers. Right answers fly. Wrong answers crash. Even in my most speculative yearnings, I still firmly believe in what I know about the universe: specifically, that its space and time scales are immense. To the questions "How big is the universe?" and "How old is it?" I confidently claim there are right answers: "Very big" and "Very old." My confidence in this flows from my belief that the countless specific knowledge claims supporting my belief in the size and age of the universe have all passed definite standards.

On the other hand, what about those yearnings themselves, the feelings that attend my reflection on those immense scales? What was the significance of Eisley's feelings when he looked at the skull of a long-dead mammal and mused to himself that "it had never lived to see humans, and I, what was it I was never going to see?" What was the significance of my feelings when I watched United Planets Cruiser C-57D flying through a field of stars? Landing on the barren fourth planet of Altair? On pain of incoherence, feelings are feelings about *something*. Yearnings are yearnings for *something*. They signify *something*. And that *something* is a thing I have to understand at least well enough to inform the feelings and the yearnings. There is an awareness here that certainly seems like a kind of knowing, but it's a vague and uncertain knowing, unlikely to suffer much reining-in and quite unlikely to pass anything like the strict standards modern philosophy would impose upon candidacy for the high office of knowledge.

It was such misgivings that prompted me to begin this book, over twenty years ago. To probe these misgivings with both flexibility and a solid anchor in my own experience, I chose to write a short novel in which Michael Howard is my stand-in. (I had a brother named Michael Howard Irvine, who died before I was born.) Use of a stand-in also allowed me to avoid straight autobiography; I focus on my own feelings because mine are the only feelings to which I have unfettered access, not because they are anything special.

Though rearranged, edited, and in some instances extrapolated, pretty much everything in the story really occurred. The story is also dramatically compressed; I could never have pulled off what Michael does in three days. The story you are about to read was long in preparation because the only way I knew to verify what it claimed was to intertwine focused analyses with unfocused feelings, relying on the feelings to verify the analyses. That took a long time. The first third of the book came together easily enough; I was posing a problem, so the job there was just to clarify that problem. But then I had to solve it. Getting the pieces to come together was slow work, but come together they did – some of the most important in just the last few years.

By compressing decades into days, I hope to have clarified both the problem and its solution well enough to make them available to you, the reader. But this clarification can only go so far, since it turns out that the *something* I am seeking to clarify is, by its very nature, unclear. More about that in the epilogue.

THE EVENING OF THE FIRST DAY

Friday, April 3, 1992

Michael drove across the Lewis Bridge of the Missouri river. In the twenty-five-hundred-mile course of the Missouri, the Lewis was the last bridge before the river surrendered itself to the Mississippi. At that confluence, six miles further downstream, thousands of tons of water quietly collided every second. Silent, swirling eddies of steel gray and muddy brown gradually entrained one another until the combination glided by the Gateway Arch of downtown St. Louis seventeen miles further downstream. For tens of thousands of years, the two rivers had been eroding a deep, wide valley into the vast mid-continent prairie. For part of that time, humans whose remote descendants would call themselves Illini had lived here, leaving their paintings on the river's bluffs and their bones in river bluff caves. Over a hundred thousand years ago, a glacier that would one day be called the Illinoian had come south and halted here, amidst a cold, barren, and forbidding landscape. Even further back in time – some sixty million years ago – an embayment of what would one day be called the Gulf of Mexico reached north along the course of what would one day be called the Mississippi to just this point, surrounded by lush forests and sandy beaches.

In the long millennia since then, much had happened. Species came and went. Mammals, emerging from the shadow of the dinosaurs, began to flourish. But for a long, long time, humans were just a subtle genetic dream; the only eyes that watched these lands answered to the souls of animals.

Now, on a warm, early spring day, Michael drove from the Lewis Bridge onto the slender peninsula of land that separates the Missouri and Mississippi rivers. Ahead on the far shore of the Mississippi he could see the small river town of Alton, Illinois. The city sits near the end of a long, twenty-mile stretch in which the river flows more eastward than southward, just after a ten-mile stretch in which the Mississippi actually reverses itself, flowing back north. Called 'the river bend' by locals, this notch is a distinctive feature of Illinois' western profile.

The day was Friday. The sky was bright blue at the zenith, devolving to a misty orange precursor of dusk at the horizon. The trees and bushes were just beginning to show a sprinkling of color. Deep purples, soft yellows, light pinks. But the broad, brown fields bordered by roads and trees still had that dusty, weathered look of late winter. The setting sun cast shadows of windbreak tree lines across the open fields. There were a few old buildings scattered about – an empty fireworks stand, a ramshackle warehouse, an old filling station – but this was basically farmland. Rich farmland, not far above river level; precarious farmland too, for the same reason. It was subject to periodic floods. Decades earlier, in the spring of 1973, Michael had returned home to Alton to find this peninsula under water. But there were no floods this year. For now, the Missouri river looked like another broad, brown field bordered by a line of trees and bluffs. For now, the river was unremarkable.

Today was Friday, April 3, 1992. Michael took in the landscape, and the feelings it evoked. He had grown up in Alton with his mother and her parents. Michael had driven this road, US Route 67, countless times. The landscape had changed very little in the decades since his graduation from Alton High School. In those decades, he had gone away to college and become first an aerospace engineer, then a scientist specializing in the physics of the Earth's rivers and oceans – the field technically known as geophysical fluid dynamics. He became a college teacher; that hadn't worked out. Now, he was an anonymous worker in a job that held no meaning for him. The surface reason for his returning to Alton was the sale of the family house, vacant since his mother's death almost a year and a half earlier. But that wasn't much of a reason; there was nothing for him to *do* in that sale. There was really no practical reason to return. The Wedge Bank had handled everything. After the death of Michael's mother, people from the bank had seen to the upkeep of the house, found a buyer, and helped close the sale. Still, Michael wanted to see the house one last time while it was still officially his family's house. This is what he told his friends back East who asked why he was flying out to the Midwest. Consciously, it was what he told himself as well.

True enough. But there was a deeper reason too. He was adrift. He had lost his way. He had lost his bearings. His sense of a place and a direction in life had weakened, and with it, the foundation for coherent day-to-day decisions. His job consisted of frequent but fundamentally insignificant crises, jerking him through an incoherent sequence of petty tasks that drained his energy and left him lurching from weekend to weekend, desperate for rest.

Except that what he needed wasn't rest, it was a revived sense of something worthy of the expenditure of his energy. Subconsciously, he was returning home to Alton, home to his house to renew a sense of belonging, of having a place in the world, a place from which to reestablish his bearings. As he looked around him now, on Route 67, it felt like every tree, every building, every turn in the road signified something – though exactly *what* was not at all clear. What was clear was that they signified *something*, and also that his renewed bearings were going to be relative to this *something*. The real task for the weekend, then, would be to get a better sense of that *something*.

The Clark Bridge over the Mississippi was immediately ahead. Just beyond it, Alton was spread out along the river and packed in among the hills and the white limestone bluffs. Hills covered with red brick buildings glowing in the deep yellow of the setting sun. Smoke stacks, railroad cars, electric power lines, and a thousand dark industrial details filled in the intricate picture. Alton was a river mill town – steel mill, paper mill, glass mill, the lead works. Michael drove up the ramp to the Clark Bridge. The rusty steel girders of the old bridge clattered by his window.

Something about the river upstream of the bridge struck him as strange. Empty…

Of course! The old Burlington Northern railroad bridge was missing! A year and a half earlier, on the first day of November 1990, Michael had watched the beginning of the demolition of that railroad bridge. He hadn't *planned* to watch it. He had no idea it was scheduled for demolition; he was stunned and briefly disoriented when it happened. He and his wife, Carolyn, were walking along the river, taking a late afternoon break from their bedside vigil of Michael's mother. Michael had been telling Carolyn about things he'd done near the riverfront as a boy: eating ice cream sandwiches while waiting for a bus by Kresge's Five and Dime; playing YMCA softball in the riverfront park; going to the movies at the old Grand Theater just up the hill. He had just shown her a small plaque commemorating the riverfront site of the Lincoln-Douglas debate of 1858 when, suddenly, puffs of smoke appeared at the joints of the bridge where the vertical girders supported the long arch of the main span. Over a hundred explosive charges erupted simultaneously. A moment later Michael and Carolyn both heard and felt a loud *"whump"* as the report of the explosions swept past them. For a brief moment, they stood mutely, unable to assimilate what had just happened; for an unsettling instant, the plain surface of the predictable world had been breached.

Several men stood patiently by a construction shack near the bridge. Unworried, their composure signaled that everything was OK. The world settled back into its reassuring rhythms. Things made sense again. The smoke cleared. The bridge remained, but now with carefully calculated gaps in its girders, one step closer to removal. Carolyn kidded Michael that while she had heard the Midwest was falling apart, she hadn't expected to actually see it happening. Unable to think of a clever comeback, he had just thumbed his nose at her.

Today, on a spring afternoon in April, a year and a half later, Michael glanced down to the dark, swirling river below, trying to remember just what the railroad bridge had looked like and thinking how large and how empty the river appeared without it. Then he looked up at the town. Dazzling flashes of light skipped methodically across the townscape as the setting sun reflected first off one window, then another. Driving down off the bridge toward Broadway, the first principal street along the river, the storefronts were already in shadow.

Turning right, Michael drove one block to the familiar intersection of Broadway and Henry, near the bottom of the Henry Street hill. Visitors expecting a flat Midwestern town were surprised by Alton's steep hills; this section of Broadway skirted the base of a big one. Henry Street plunged a hundred feet from the hilltop, crossed Broadway, and then descended one block further to Front Street at river level. Numbered streets up the hill from Broadway (which had originally been called Second Street) valiantly tried to run parallel to it; Third Street lay immediately above Broadway; Fourth Street was still further up the hill, and so on. But – as any high school geometry student knows – parallel lines only really stay parallel on a flat plane. On a curved surface, they paradoxically tend to run into one another. Here, at this intersection, the hill turned away from the river. Veering twenty degrees to match the hill, Broadway crossed Henry Street. A block further along, Third Street pitched down to end its course, intercepting Broadway at an acute angle. The building on this triangular block was aptly named the Wedge Building. The wedge looked like the prow of a brick merchant ship plowing through a concrete river, with Broadway off its starboard bow and the rising wave of Third Street off its port. The primary business in the building was the Wedge Bank, the executor of his mother's estate. There was also a small restaurant in the building called the Wedge Cafe, at the corner of Broadway and Henry. The cafe caught Michael's attention as he entered the intersection. He wasn't really hungry, but figured he eventually had to eat. Now was as good a time as any. He pulled off Broadway just beyond Henry and parked.

Michael got out, stood up, and stretched, realizing how stiff and tired the trip had left him. His flight from Baltimore had been delayed because of equipment problems. However, since there was really nothing to be done in Alton that night, he felt unhurried and had spent the three-hour delay watching planes and reading a book. Still, by the time he arrived in St. Louis, picked up his rental car and drove to Alton, he was surprisingly drained. He closed the car door, crossed the street, and entered the cafe. Inside, the room appeared dark at first; a row of windows to the left were shaded by slightly opened blinds, casting a bright array of thin lines on the floor and the far wall, accentuating the surrounding darkness. The room, like the building, was triangular. One table was occupied by three gray-haired women; the rest of the tables were empty. A waitress stood behind a lunch counter opposite the door, smiling at Michael as he entered. He sat down with his back to the corner that matched the angular meeting of Broadway and Henry. He picked up the menu – headed with the prominent banner, "Open 24 Hours" – and perfunctorily scanned it as the waitress approached. He ordered his standard away-from-home dinner: hamburger, fries, and coffee.

Michael picked up a copy of *The Telegraph* – thirty years ago it had been *The Alton Evening Telegraph* – that had been left on the chair next to him, and browsed the headlines: "Agreement means extra five deputies for Godfrey"; "Officials act to put an end to vandalism at Alton High"; "Proposed bill takes aim at Laclede Steel." A small box contained a note: "Just a Reminder: Central Daylight Time Begins Sunday." Opening the paper, Michael saw an ad for Piasa Lincoln-Mercury. He smiled to himself; now he was *really* home.

The word "Piasa" meant something like devil bird in the language of the Illini – earlier inhabitants of this region. In 1673, the French Jesuit missionary Jacques Marquette had first come to this stretch of the Mississippi and, as he wrote later in his journal, "while skirting some rocks, which by their height and length inspired awe, we saw upon one of them a painted monster which at first made us afraid, and upon which the boldest savages dare not long rest their eyes": the mythical Piasa. Every school child in Alton learned some version of the legend of Ouatogo, the Illini chief who killed the Piasa. Ouatogo camped alone for a month of fasting and prayer, pleading that his people be spared the Piasa's devastating attacks. On the last day of his fast, the Great Spirit appeared and directed Ouatogo to select twenty of his warriors, arming each with a bow and a poisoned arrow. Each of the twenty was to hide in a designated spot. Another warrior was to stand in the open, offering himself as a willing sacrifice: Ouatogo. When preparations were complete, Ouatogo stood alone, unarmed and in clear sight of the bluffs. The Piasa appeared and stood motionless for a long time at the crest of the bluff, studying its victim. Then, abruptly airborne, the bird dived. Ouatogo closed his eyes and began his death chant. The sound of twenty arrows. A terrible, alien scream. The sudden impact of wind as the mortally wounded bird swept over Ouatogo, trying desperately to regain altitude. Ouatogo opened his eyes and turned in time to see the bird crash into the river and disappear.

The original painting of the Piasa seen by Marquette had long since weathered away. In the nineteenth century, an artist from Alton had painted an image of the Piasa on the bluffs just west of town, a few miles upstream of the future site of the Clark Bridge. As that painting weathered, it too was followed by another rendition. Then another. Finally, a metal version was secured to the cliff in 1983, only to be removed later because of the problems gawking tourists caused for the river barge company at the base of the cliff. The symbolic force of the painting had weathered as well. This particular version of sacrifice supported by a supernatural spirit had been replaced by a variety of different ones; here, in Alton, those symbols were largely Christian.

No longer a supernatural symbol, the Piasa bird had survived as a civic symbol. The city of Alton was filled with Piasa-this and Piasa-that: Piasa Cab Company, Piasa Photo Service, Piasa Insurance, Piasa Motor Fuels. Michael had opened his first savings account with Piasa First Federal Savings and Loan – which had gone under with other S&Ls in the eighties. In fact, an easy way to identify a real resident of the city was to have them pronounce "Piasa"; anyone from Alton would confidently say PIE-a-saw. As a student at Horace Mann Elementary, Michael had come up with  + a +  in a rebus contest. And was highly impressed with himself. For days. When confronted with the word "Piasa," non-natives would usually come up with something that rhymed with "plaza," like pie-AZ-ah. Looking at the word now, though, Michael thought to himself that pie-AZ-uh seemed a more plausible pronunciation than PIE-a-saw. Did that make him a non-native? Maybe. After all, he was a native of the Alton of the '50s and '60s, not of the '90s.

Michael continued to look through his *Telegraph*. He found himself paying more attention to the ads than the articles. Silently, he spoke the names to himself: "Piasa Lincoln Mercury, 2350 Homer Adams Parkway"; "Duke Bakery, 819 Henry Street"; "Elias-Smith Funeral Home, 2521 Edwards Street"; "Alton Refrigeration and Electronics, 'Your Home-Owned Appliance & Electronics Store – One Block East of the Clark Bridge', 550 East Broadway." *That* was just across the street from the Wedge Cafe.

Before long, his dinner arrived. The elderly women were talking in subdued voices. Sporadic conversation came from the kitchen. Otherwise, it was quiet. The occasional clink of his coffee cup and the rustle of paper as he read the *Telegraph* only accentuated the silence. The bright array of sunlight on the far wall, filtered by the blinds, slowly ascended as the sun set in the west. Slowly, very slowly, the thin pattern crept further up the wall, growing dimmer as the diminishing angle effectively closed the blinds. After the moving sun and the fixed blinds had conspired with geometry to extinguish the direct sunlight on the wall, the room seemed marginally brighter to human eyes. Michael absently noted the oddness of a room becoming apparently brighter when it had actually become darker.

Finally, finished with dinner, he sighed, stood up, and walked to the cash register. The waitress pushed the green check into a slot in the register, punched in the amount, and waited for the machine to grind out the total and shove open the cash drawer. She gave Michael his change, withdrew the check and poked it onto the spike with the rest of the day's checks. Those vertical spikes always made Michael uneasy.

Standing on the sidewalk by the cafe, waiting to cross the street, he looked around. The scene was quiet, almost somber; the Friday evening rush hour didn't have much of a rush to it. Some of the stores near the cafe were empty; storefronts were boarded and painted a uniform gray. The cafe itself would cut back within the year, going from twenty-four-hour service to just breakfast and lunch – five in the morning till two in the afternoon. While Alton had been on high enough ground to escape serious damage from the periodic Mississippi floods, it had already been swept under by more subtle, economic ones. The mills along the river were mostly obsolescent, part of what was picturesquely called the Rust Belt.

Back in the car, Michael continued east along Broadway. In a few blocks, he passed Springman Lumber Company, where his mother had worked as a bookkeeper for thirty-three years. Despite Alton's depressed economy, the Springmans were still in business. They just weren't going to give up. Next came the Owens Illinois glass works. Much of the old plant was gone. The remaining buildings were now far back from the road, separated from Broadway by a large, empty field. As Michael drove along Broadway, changing patterns of light and shadow played among the idle industrial shapes, bright yellows against grays, browns, blues, and rust reds. Empty, silent stretches that had once been filled with noisy industry. Beyond Owens Illinois were the massive grounds of Laclede

Steel. Here, at the steel mill, Michael turned right onto Cut Street, heading back toward the river that these industries energetically consumed. The street threaded through acres of railroad tracks and corrugated metal buildings. There seemed to be nobody around, though lights dotted some of the buildings and low, rolling trails of smoke showed that something was going on somewhere.

Michael continued on through Laclede to the end of the road, and stopped in front of the entrance to what had once been the Alton Box Board Company. In 1939 his grandfather, a manager with the Box Board, suffered a serious accident: his right hand was crushed between two enormous steel rollers. After many grafts and much care (the family had bought presents for the nurses who diligently attended him) he regained limited use of his hand; his handwriting became legible again, but remained clumsy. The dark menacing shapes inside the paper mill had awed the child Michael; it had seemed fantastic to him that something as delicate as paper could come from such a massive, blustering cauldron of steel and chemicals. A decade later, as a young engineering student, he had felt much more comfortable with industry, confident that he understood the basic chemistry and physics that constituted the backbone of these places. Now, though, the wholly adult Michael quietly studied the obscure silhouettes inside the large door of a nearby loading dock. Neither awed nor comfortable, he now sensed something else. He tried to get some grasp on the unfocused but persistent feelings attached to this place. Unsuccessfully.

The final minutes of the sunset were magnificent but a bit ghostly, amid these decayed industrial profiles. It was as if the buildings were consciously brooding on the dying colors of the western sky, with sightless window-eyes and silent, gaping doorways. The roofs and towers and smokestacks of the paper mill and adjacent steel mill – everything above a very distinct level – were bathed in the gleaming yellow stream of the setting sun. Below that plane, dark blue shadows. The mills seemed to be sinking into a blue-black lake of mist and smoke, ineffectually punctuated with tiny white lights. Something in this stark scene, at once familiar and foreboding, attracted Michael, compelling him to stare at every detail. In the distance, a diesel yard engine was slowly pulling a long line of coal cars out of a low brick building lined with translucent windows, backlit by a greenish-white glow. Michael looked more closely at the switch engine itself and finally made out an engineer, the first person he had seen in either mill.

He rolled down the window, to see if he could hear anything. A deep rumble, perhaps. The shearing sound of steel on steel. *Something*. But there was only the wind and a few distant, indistinct sounds. Restarting the engine, filling the car with the sounds he could not find outside, Michael turned around and drove back up Cut Street, wondering to himself how many times Grandpa had taken this road, and what he had been thinking and feeling. Back at Broadway, Cut Street changed its name to Main Street and, like almost every street coming from the river that continued beyond

Broadway, ascended steeply. Getting up this hill on an icy winter day must have been a real adventure; Michael imagined Grandpa struggling up Main Street hill in his Studebaker.

As he reached the crest, dusk was settling around him. Street lights were starting to come on. Driving along Main, he glanced into the front windows of the houses; he could make out table lamps, wall lamps, ceiling lamps. Occasionally, a dim, flickering bluish glow marked a room lit solely by a TV. More often, though, it seemed as if a solitary light or two held back the darkness. Most of the windows of most of the houses were dark.

Odd. Even houses with lighted rooms looked to Michael like caverns of shadows. A back room – a kitchen, say – might be brightly lit, but when viewed from the street outside for an instant through the front window, the immediate impression was more of the dark, heavily shadowed front rooms than the bright back room. It is a simple fact that in the evening, people live in houses with more darkened rooms than lighted ones. These people don't think about the fact that their houses are full of shadows, but the shadows are there nonetheless. Michael observed this simple fact as he drove.

He passed Main Street Methodist Church, on his left. His grandmother had been a regular member; his mother, who was married there, had later attended out of a sense of duty to Michael, thinking it good to expose her son to religion. As a pre-teen, Michael had become quite active in the church. Shortly after becoming a teenager, though, he stopped attending; what he heard in church didn't make sense to him and after all, as a future engineer and scientist, he took religion to be his natural enemy. Mom and Grandma were distressed; Grandpa, who never went to church if he could avoid it, quietly approved.

For a short time, Michael drove around upper Alton, turning here and there as the whim took him. Finally, he arrived at the house: 1918 Washington Avenue. Brick, two stories, with a full front porch, a detached wooden garage and a small yard, the house had been built in 1934 for a family who soon lost it to the Depression. Repossessed by Springman Lumber, who had built it, the house was then rented in 1936 to Michael's grandparents. Ten years later, his grandfather – who had grown up on a small Ohio farm at the beginning of the twentieth century, and never thought he'd own a house in a town – bought it for the sum of $2,300. The front of the empty house was starkly illuminated by a single streetlight. A table lamp in the living room, run by a timer from seven to ten-thirty every evening, was clearly visible through the closed curtains. Michael unlocked the front door, stepped in, and put his suitcase on the floor by the couch. The smells were stale but very familiar. He turned on the overhead living room light and closed the door. Most of the furniture was still in place. The new owner would take it along with the house. An 'as is' sale.

He walked absently through the living room and dining room into the kitchen, turning on lights as he went. He opened the refrigerator; it was empty except for a few Cokes, a bottle of ketchup, and an unopened six-pack of grapefruit juice. At the prod of a vague memory, he opened the freezer. Yup, there was that pint of vanilla ice cream he'd bought, wondering if he'd ever get to it. He hadn't. And felt foolish now, having bought it. Closing the freezer, he pulled out a Coke and returned to the living room. His memory still judged the rooms with a child's sense of size and proportion. As an adult, he always felt just a little awkward. Too big. As if just a few moments ago he had been seven years old sitting in a big chair, then suddenly found himself on a small chair in a shrunken room. Mark Twain, returning home late in life to Hannibal, Missouri (not far upriver from Alton), had said, "It all seems so small to me. I suppose if I should come back here ten years from now it would be the size of a birdhouse."

The house was quiet. Michael closed his eyes and leaned back. Next to him, the electric timer that controlled the living room light hummed. It reminded him of the clock on the wall in his mother's hospital room. It was the last room and the last clock she would ever see.

Tuesday, October 23, 1990

CAROLYN WOKE MICHAEL just before dawn. "Dear, you have a call from Alton. It's your mother."

He took the phone, groggy but a little anxious; there was no reason he could think of to be getting a call from Mom at this hour, unless something was wrong. Something was wrong. His mother's tone was anxious. She had come to the Alton Memorial Hospital's emergency room early that morning with a sharp pain in her lower back. Nobody could find the cause. She had gotten worse, her white blood cell count soared, and now the doctor was insisting on surgery. She was hesitant; she wanted her son's opinion. But after a brief conversation, a no-nonsense doctor took the phone; he told Michael his mother's colon was perforated and that surgery was absolutely necessary. Michael asked a few unfocused questions; the doctor politely but firmly cut him off: "Mr. Howard, I have to check on the anesthesiologist now. Please give a phone number where we can reach you today to the nurse I'm about to put on the line." He could overhear the doctor telling the nurse that the man on the line was the son of Virginia Howard who was now being taken to operating room number three.

Monday, October 29, 1990

A WEEK LATER, AFTER successful surgery and a deceptively swift release from urgent care, Michael's mother was sitting up in her hospital bed, going over plans. Michael would take care of a few items, drive her car back to the house, and take a limousine to St. Louis airport in time to catch the return flight to Baltimore. She would have Roy Greywood, an old friend of the family, keep an eye on the house until she could go home. She insisted that she'd still be coming out to Baltimore for Thanksgiving. Michael thought that unlikely but, like his mother, he was still not facing the consequences of her emergency surgery. A warning tremor had just shaken her life, but neither of them was willing to think about the possibility of impending death.

He looked at the clock on the wall. It was two-thirty; he didn't have to leave for half an hour yet. It was quiet. Unlike the urgent care room, crammed with noisy equipment and opening directly onto a room of busy doctors and nurses, this was a regular hospital room which, for the time being, his mother had to herself. The electric clock on the wall hummed steadily. He continued to look at the clock.

There was something strangely unsettling about it. The face of a clock is rich in associations: two-thirty means that the afternoon is about half over, and the adult workday is well along; two-thirty means that there is only an hour of school left; two-thirty means that there is still plenty of time till dinner. Plenty of time left in the day. All this in the look of the hands – the minute hand pointing straight down, the hour hand slightly above horizontal, as if holding open the last memory of the morning. But here, now, it seemed alien, unlike any other two-thirty. This was time disconnected from human habits. No work, no school, nothing of the business of a human day. This was not a domesticated two-thirty. This was two-thirty on the clock of the universe. The idea of even calling this 'two-thirty' – whatever 'this' was – began to feel absurd, a hopelessly inadequate marker buoy pretending to delimit unlimited time.

They both sat silently for a while. Michael suggested that they play a few hands of gin rummy till it was time for him to leave. His mother fidgeted, trying to remember where the cards were. She asked Michael to check the bedside table, guessing the cards were in the drawer; her guess was correct. They played a few hands of gin rummy. They had played cards for as long has he could remember; it was a good way of spending time together. Finally, though, three o'clock approached. He went through the bills one last time; he'd pay the electric bill that was due, and Roy would bring her mail to her as long as she was in the hospital. She said she'd see him in several weeks for Thanksgiving. He nodded.

"OK, Mom. I guess that's about it. I'll see you later. I love you."

"I love you too, dear. Have a safe trip."

He kissed her and walked out the door. As he left, he glanced at the clock; it was precisely three.

Virginia watched him go and sighed. She was sighing a lot these days. If only Michael weren't so withdrawn, so undemonstrative. If only he lived closer! She sighed. Her anger at being left alone subsided as quickly as it had arisen, replaced by more habitual self-recrimination: if only Michael hadn't been an only child; if only she'd remarried. But what would people have said? Sighing yet again, she laid her head back, closed her eyes and deliberately tried to stop that familiar train of thought. A memory appeared: the memory of how she had felt in 1953 sitting in a railroad passenger car, next to Michael, returning to Alton following the separation from her husband, Guy. She had been worried to death over what Guy might do, worried about Michael who had reverted to sucking his thumb, worried about finding work, worried about a thousand other things. But, somehow, she was worried most about what people would say. She remembered her first thought getting off the train: "My God, Dad's taken off work!" An instant later, she realized that Helen Davis, a friend who had been one of her bridesmaids, was standing a few feet behind Dad: "Oh my God! Helen! Now everybody is going to know I've left Guy!" Decades later, people would tell her that she was very courageous to do what she did; most women in the early fifties would have stayed put. She would listen to their praise and even agree in a certain detached way. Yet she hadn't *felt* at all courageous. Instead, she felt lucky to have a father and mother who would support her. And she felt guilty to have to take that support.

Now, looking around her hospital room, she wondered what to do with herself to get her mind off these tired but still troubling old memories. Since she was alone in the room, she decided she could watch TV without bothering anyone. Watching with little interest, she was soon asleep.

Spring, 1952

A warm day. The clouds and drizzle of the weekend had moved out, replaced by dryer, more comfortable air. Virginia Howard took her son outside and headed for the field east of the house. Actually, there were fields in every direction. By the end of the century, there would be a shopping center where the house now stood. But now, this was Downers Grove Rural Route 2, southwest of Chicago. Still mostly farmland, with a few houses intruding – most new, built after World War II.

Guy was gone for the week to a business convention downstate in Springfield. Things were quiet, now that he was gone. Peaceful. Virginia held her son's hand, carefully looking both ways at the road – not that there was any traffic – and walked quickly across the road into the field. A small brook cut diagonally through the field, a brook rippling over small rocks, running alongside new grass. Virginia had brought one of Michael's old water color sets. The small, square paint containers were easily detached and would bob entertainingly in the brook, weaving in and out through the rocks, riding the small waves. Michael absolutely loved this, consumed by the unbounded amazement of a three-year-old child. They reached the brook and, of course, Michael wanted to launch the paint-boats himself. The first one, a bright yellow, landed smack in the middle of the brook and zipped off, not stopping for several yards. Michael ran to follow it, splashing into the brook; Virginia promptly pulled him out. Returned safely to shore, he continued to launch one paint-boat after another. A red one. A blue one. A green one. Wizard – a speckled black and orange cat – appeared and started to bat at the moving targets. Michael was delighted. A squirrel in a nearby tree fussed, evidently less delighted at the cat's appearance.

Virginia set up a folding chair near the brook, absently browsing a magazine while keeping a weather eye on Michael. It was such a beautiful day, and it was so calming to be able to just sit outside, away from the tension, away from the craziness, and just take in the scene. Michael enjoyed playing by himself and sent his little boats again and again down the brook, to a constantly changing story-line. First, they were boats for the ants. Then they were boats for the cows – several black and white cows were visible in the field behind the house. Then it was silly to think of a cow in the boat. The boat was too small! Then, the boats contained little boys. Good little boys. Bad little boys. She gently tried to nudge the story-line in positive directions. Michael was really in a world of his own today. Happier than she had seen him in some time.

The day before, as Guy was getting ready to leave, she had played one of Guy's new records, at Michael's urging. Guy had marched out of the bedroom, grabbed the record from the phonograph, and smashed it to pieces, right in front of Michael. Michael hadn't cried. He hadn't made a sound.

Virginia knew, at some level, that the situation with Guy couldn't go on. He was becoming increasingly erratic. Sometimes he was full of exaggerated love; sometimes, hostile and brooding. Recently, however, his hostility was taking an increasingly cold edge. It was no longer simply connected to an episode of anger triggered by a clear provocation, however farfetched. It could erupt for no evident reason at all.

That record. What conceivable provocation was that record? When Michael picked the record, she thought Guy would be happy to hear his son listening to his father's music. A nice send-off. A demonstration of affection and interest. Why? Why?? At first, Michael had been quiet. She wanted to talk to him, but decided that any damage control would have to wait for Guy's departure. Once he was gone, Michael had been extremely demanding of attention. He followed her wherever she went. She had stepped outside once to take down some laundry. Michael had not seen her leave and, when he couldn't find her, screamed in panic.

But today, Michael really seemed to be happy. And it was so relaxing, so healing to just sit out here, reading, watching clouds. There was really no serious trouble Michael could get into out here, so she didn't feel she had to monitor him closely. After a while, though, she could hear signs of fatigue, and decided it was time for his afternoon nap. A decision Michael instantly and vocally protested. She hauled him, crying, back across the road into the front yard. Wizard had seen where they were going and was already at the front door, waiting with typical cat impatience to get in and get at the food. She took quick advantage of the situation, pointing out to Michael that Wizard wanted to go inside. Michael stopped crying for a moment and looked at the cat, who rubbed up against the doorway, tail erect. Virginia now had to restrain her son from rushing headlong at the house he had been desperately resisting a moment earlier. She reached the front door, managing to keep Michael sufficiently in tow so as not to spook the cat. She opened the door. Wizard lunged inside, immediately followed by young Michael.

Wednesday, October 31, 1990

HALLOWEEN. BACK ON the East Coast, at a noontime company party, Michael sipped a Coke and then tried to eat his crumbling cake without spilling too many crumbs on the floor. Some people had come to work dressed up in costumes, but not Michael. It would have made him feel silly. He preferred to move back to the edge of things and just watch. Actually, he liked Halloween. He thought about a Halloween parade back in Alton when he was in fourth grade. He, his mother, and her parents had gone to Springman Lumber to get a good view of the parade up Broadway. Sputnik had been launched, and the adult world had been catapulted into the space age, the excitement of a new frontier subdued by the disquiet of a cold war. Halloween was a bit spookier than usual. Oblivious to such larger concerns, Michael was fascinated by the empty store and quickly slipped off into the dark warehouse, eagerly taking in the atmosphere: the long piles of lumber, the concrete floor, and most of all the strange mixture of smells: sawdust, oil, glue. He leaned down and sniffed a sheet of plywood. Why did it smell this way? Michael's grandfather disrupted these musings, exasperated: "Come on out here, sonny!

We come to see the parade, not to snoop around Springman's store." Years later, Michael's foremost memory of the parade itself was from the back of the store, looking toward the front: Grandpa, Grandma, and Mom by the store window, the heads of people outside, brightly-lighted floats moving by. The usual ghosts and goblins were joined this year by a smattering of Sputniks. But all this was somehow surreal, strangely framed by the darkness of the store itself.

Back in the present, Michael's thoughts turned to his mother at Alton Memorial Hospital. Things weren't good. Tuesday, the morning after he returned to Maryland, he'd gotten a call from Roy Greywood. Virginia had awoken at three in the morning, totally disoriented and frantic. Roy had been called to the hospital to sit with her till dawn. With sedation, she seemed to improve, so he had returned to his house and called Michael. "I think she'll be better in the daytime, Michael." She wasn't; pneumonia set in. Now, standing alone, unaware of the party around him, he agonized about what to do. Fly home? *Again?* This was getting expensive. And if he went home, what would he do? How long could he stay?

The department secretary came up to him, a concerned look on her face. "Michael? There's a call for you from . . ." she paused to check a note, ". . . from Alton Memorial Hospital."

He raced to the phone. A nurse was on the line; she spoke very softly. “Mr. Howard? I think you’d better come on out here, as soon as you can.” Michael immediately called Carolyn. He would fly to St. Louis at once and take a limo to Alton; she would follow on Thursday morning, after taking care of things at home.

Friday, November 2, 1990

Michael and Carolyn were sleeping at the family house, in Michael's old upstairs bedroom. The phone rang. Michael opened his eyes, briefly confused about where he was. Carolyn was the first to move, running downstairs to take the call. A minute later she called up, "Michael, it's the hospital. Your mother isn't doing well and the nurse thinks we should come right away." They dressed and walked outside. It was nearly two in the morning. There was a full moon. The leaves crunched noisily as they walked to his mother's car. Ten minutes later – the hospital was only half a mile away – they walked the now-familiar route to his mother's room. The walls were still decorated with Halloween figures: pumpkins, black cats, ghosts, skeletons. He commented to Carolyn how little some of these decorations had changed in the last thirty years. When at last he could see his mother's room he noted with puzzlement that the door was partly closed and that the room was full of people. As he and Carolyn approached, a nurse emerged, diverting them to the adjacent room.

Code Blue. Mrs. Howard had requested that extraordinary measures not be taken. Did he want to see her?

He went into his mother's room. The people had all left. It was quiet and dark, except for a single light above the bed. All the equipment that had surrounded her had been removed. The room looked empty. His mother lay in the bed, her eyes closed. Michael felt something inside himself give way.

THE NIGHT OF THE FIRST DAY

Friday, April 3, 1992

Now, a year and a half later, Michael sat quietly in the living room. He thought back to the night of his mother's death: driving back from the hospital; a sense of disorientation; breaking down in the kitchen, crying violently, with Carolyn holding him; both of them walking uptown to the Village Inn a few blocks away at four in the morning; drinking coffee and eating a piece of pie; returning to the house to begin making arrangements with the funeral home and the Wedge Bank's trust department; lying down for a nap and dropping into an exhausted, dreamless sleep. It was Friday, November the second – his wedding anniversary. The front page of the *Telegraph* had a photograph of the demolition of the railroad bridge he and Carolyn had witnessed Thursday. The headline read "Alton Memorial plans $12 million expansion." The obituary inside the *Telegraph* stated that Michael's mother, Mrs. Virginia M. Howard, had died at 2:09 a.m. At Alton Memorial.

2:09 a.m.? How did they arrive at such a precise time? Michael imagined the clock in her room, the one that had shown two-thirty a few days earlier.

He imagined the emergency staff working beneath that clock. Perhaps a nurse had looked at that very clock when it was clear that Mrs. Virginia M. Howard had died, to determine that it was precisely 2:09 a.m. Simple enough. And yet . . . What is time, after all? Just an independent coordinate that, along with its three spatial siblings, serves as an objective index for all the events of the universe? Events like birth and death?

. . . no, it must be something more. Time is something more. Time means something more. But what?

Michael stared blankly at the lamp beside him. A dull thud from the kitchen sent a chill down his back. Then he realized it was only the refrigerator, cycling on. After all, it hadn't been opened for a long time. The familiar, mechanical sound of the compressor should have been reassuring. Instead, it was unnerving; it seemed to heighten the silence. From deeper in the house, another sound. This time it was the furnace, triggered by the night air that had swept in with Michael. The house, disturbed, was trying to reestablish itself. To assert himself in turn, he stood up and wandered through the first floor, turning on lights and aimlessly opening and checking this and that. It didn't work; the house continued to murmur and the back of his neck was getting clammy. Finally, he laughed out loud, "OK, house, I'll leave you alone for a while. I feel like going for a walk anyway."

Michael put the cap back on his unfinished Coke and walked to the kitchen, putting the Coke back in the refrigerator. Closing the refrigerator door, he walked past the kitchen table to the basement door. The basement was off to the left, the door to the back yard, to the right; he opened that door and stepped outside. In the back yard he walked a few yards to the old, white, one-car garage, desperately in need of paint. The two garage doors were so badly warped they would not close fully; a large padlock and chain hung uselessly from the handle of one. Michael tugged at the door, which finally scraped open along the concrete driveway. The inside was empty, of course; his uncle had taken the car after the funeral. Decades of car oil were still evident on the floor.

Closing the door, he turned around and wandered down the driveway. Coming to the street, he crossed, angling to the right, heading from old habit toward the house of Bernie Kohler, his first real friend. Bernie's father had been a high school classmate of Michael's uncle, one of his mother's brothers. Later Bernie's father had bought a house at 1821 Washington, a block south of 1918. Bernie was a year older than Michael, who had therefore seen Bernie as sort of a big brother. Over the years, though, they had drifted apart; after two decades, they met again – at the funeral of Michael's mother.

Now, turning off Washington onto Judson, Michael walked near the site of Bernie's house; the house itself had been torn down to make way for an Illinois Bell Telephone building. Judson was on the route he had taken to Horace Mann Elementary and now, without any conscious plan to go anywhere, he automatically retraced that route and arrived shortly at the two-story brick school building. Replacing a smaller brick structure on the same site, this building had been completed in 1955 to accommodate the post-war baby boom. And it really *looked* like a school building. Michael wondered occasionally if there was some fabulously successful architect in the early 50s who had sold everyone on the idea that elementary schools ought to have a single row of narrow rectangular windows topped by an expanse of translucent glass bricks; maybe the idea was to give the kids plenty of light but not to actually let them see much. Might distract them. Anyway, it seemed that almost every small town Michael had ever come across had a school with just this same style; he'd muse to himself, "Damn, that looks just like Horace Mann!" He walked down a narrow path along the north wall. The playground behind the school was now blacktop, illumined by a harsh blue-white floodlight atop the gymnasium; crossing the playground, Michael sat down on the stairway to the gym, shielded from the direct glare of the floodlight. Leaning against the wall, he looked back at the north wing he'd just circled and remembered a conversation he'd had from just about this position in the spring of 1955, on the back stairway of the old school building – the gym and part of the east wing had not yet been built. Having just graduated from the first grade, he and a classmate had been sitting on the old stairway looking over at the new building and guessing which window was the principal's office. The following fall, he discovered that the room he'd guessed was none other than his second grade classroom.

A man came into that very room, backing the door open and pulling something after him as he switched on the overhead light. Evidently a janitor. Involuntarily, Michael thought of his father, Guy Howard. A bright, creative man, Guy was diagnosed paranoid schizophrenic and spent the mid 50s in a mental hospital established in 1872 as the Northern Illinois Hospital and Asylum for the Insane. After his release, still unable to get along with most people, living in a painfully detached world, he spent much of his life as a school janitor, working alone at night. Michael's mother left her husband in the early summer of 1953; he was committed late the following fall. She would later say she made the decision when Michael said to her, "Mommy, can we get a new Daddy?" With her son in tow, she came back downstate from Chicago to her parent's house in Alton. A month later, Guy showed up, threatening Virginia in her father's front room. Her father, armed with a pistol, walked into the front room and calmly told Guy to leave and never return. What shocked everyone was that pistol; nobody knew that Virginia's father even owned one! Guy left. Shortly thereafter, he threatened a Chicago policeman with a rifle, was arrested and involuntarily committed to the renamed Elgin State Hospital. That fall, five-year-old Michael began kindergarten at the old Horace Mann.

Now, the forty-three-year old Michael watched this unknown janitor cleaning that same second grade classroom. The janitor seemed to be picking things up. Mostly, though, Michael couldn't really see what he was doing, except for a few minutes when the janitor cleaned the green 'blackboard'. An old grade-school board with guidelines for writing letters. First, he erased the day's chalked letters, then wiped the now-empty lines with a damp cloth, turning the light green into a shiny, dark green on which the white lines stood out clearly. Shortly thereafter, the janitor finished and left, switching off the light and closing the door behind him. The only remaining light was that in the hallway shining through the single window in the room's door. Michael sighed. And sighed again and stood up abruptly. Crossing the blacktop playground, he strode back up the narrow path to the street, suppressing any further thoughts about his father.

It was just after eight o'clock. Half a mile away, Amtrak 305 – the Ann Rutledge, northbound daily from Kansas City and St. Louis to Chicago – approached Alton Station, sounding its horn. Michael listened as he walked. That low, moaning sound was a mnemonic marker for Michael, whose conscious memory of his first years in Alton was empty of anything outwardly related to the wrenching events of the divorce; nearly twenty years passed before Michael was told the story of his Grandfather's armed ultimatum in the living room, and Guy's subsequent committal. But Michael's subconscious mind carried the memory of that time, recorded in images from the world of a five-year-old.

The one clear seepage from unconscious to conscious memory was a remembered dream, keyed to that train whistle: a dream of the local A&P grocery store on Washington Avenue, lighted from above by a single, harsh street light. In the dream the store was being destroyed. Yet, paradoxically, there was nothing moving in the dream; there was only that one street light, surrounded by dark night. Somehow, the menace was in that sound. The sound of the train whistle.

Michael wandered along streets and alleys for a long time, alternately evoking and suppressing memories, trying to sort out conflicting, vague feelings. Circling blocks, looking at houses, he probed for . . . something. At times near tears, he wondered if this was nothing more than simple grief, mixed with the usual regrets – this not said, that not done. Yet it was clearly more. A deep disquiet. A ravenous hunger. A compulsion to make sense of things, to think through and clarify. To see something. To bring into focus something inside him. To express something. To find words for this mysterious significance – this mysterious *something* – that seemed to hide under every stone, to hang in the humid night air.

Eventually, though, fatigue began to subdue these broodings. Sarcastic little comments began to pop into his head: "Get the hell inside, Michael"; "For God's sake, man, you're tired!"; "Go to bed, already!" Finally, at the crest of a hill that descended sharply to Alton High School, he consciously decided it was time to go home; decision made, both the internal brooding and the sarcasm quickly subsided as he turned around and walked two short blocks, through a small alley, and in the back door.

Walking back to the front room, turning out lights as he went, he picked up his suitcase by the front door and walked up the front stairs, bending to avoid the low ceiling. Michael's mother and grandparents had all been relatively short; Michael was tall, like his father. The room at the top of the stairs had been his bedroom when his mother and he first arrived in Alton following the separation. There was no bed here now, just a chest of drawers and an old drawing table. On one side of the room, three windows opened onto the street in front of the house. On the opposite side, two doors led to the two back bedrooms. The second one, further from the stairs, had been his mother's; the first, his when he had gotten old enough that a private bedroom was in order, one that his mother didn't have to walk through to get to hers.

Michael entered this, his second bedroom. Putting the suitcase down on a chair, he unpacked his shirts and pants. He walked to the closet door, opened it and switched on the small light bulb fixed to the low ceiling. He hung up his clothes. A few feet away, at the far end of the closet, stood the door to the attic. A frayed cord from the attic ran along the floor and up the wall to the closet lamp, ending in a large plug that hung alongside the light fixture. Michael plugged it in; the light behind the attic door came on. The attic had always been a bit creepy, full of boxes of mysterious stuff from the past. As a child, he had always enjoyed exploring the attic – even if it was a bit creepy. The attic door was held in place by two small nails that Michael bent aside. He kneeled down and opened the door. The attic was empty now. The light showed the sloping beams of the roof that came down to meet the floor along one side. The dusty smells and the silence. *So familiar*! Soon, though, his cramped stance became uncomfortable; he replaced the attic door, stood up, and stepped back out of the closet. Closed the bedroom blinds and donned pajamas.

Still restless, he wandered back to the head of the stairs and sat down on the top step. The low rumble of the furnace dropped away; the house, satisfied now with the temperature, slipped back into silence. Michael stared vacantly at the banister. The stairway descended three steps to a small landing, swung left and continued down to the front door, next to the living room. A single window above the landing was covered by a white shade, cracked and dusty – brittle with age. A dirty net of cobwebs draped from one corner of the window sill to the adjacent wall. He stretched out and leaned back, hands behind his head, studying the banister. The railing, the vertical supports, and the caps were all painted white. The stairs themselves were varnished natural wood, covered by a runner of thick, green carpet. The walls were a lighter shade of green; the stairwell, a dirty white. Except for an occasional car passing outside, the house was quiet. In a dark brass fixture above the landing window, a single, uncovered bulb immersed the stairwell in a bright light.

A little too bright, Michael thought. Funny, though – the naked bulb in the old fixture ought to look awkward but it didn't, really. If this had been in some other house, he'd have judged the fixture unfinished, needing some sort of cover to tone down and distribute the light more softly, more evenly. But here it wasn't just a light, it was the-light-in-this-stairwell, and this is how it looked. It had always looked this way. This was how it was *supposed* to look. To be sure, he couldn't exactly remember ever looking at the light before now, yet he knew he must have seen it just about every time he had come up or gone down these stairs, until every feature, every detail of the light fixture was invested with the associated, projected memories of everything he had been thinking or feeling. How many times had he gone up and down these stairs? Twenty thousand? Thirty thousand? The sheer magnitude of the experience in terms of trip-counts completely exceeded meaningful grasp. Yet all this was packed into memory, an inconceivably long thread coiled down an unthinkably deep well, into which he peered, now, from the top of the stairs.

He closed his eyes. The quiet and something in the smell of the stairs – old dust, maybe – brought back a memory of sitting here as a child, the bedroom behind him dark, the living room below him bright and alive with adult conversations; his aunt and uncle from the East Coast had been visiting. He could remember being able to pick out each of the voices, and sometimes who was talking to whom. He was supposed to have been in bed, but far too excited to sleep, he had worked his way down the stairs, a step at a time. Patiently, he had sat quietly on each step till he got up the courage to move down one more. He had concentrated on looking mostly at his own feet on the step below him, only furtively glancing over to the assembled adults, thinking that this somehow minimized his offense: if he didn't look (much) then he wasn't really breaking the rules. And if he couldn't see them, maybe they couldn't see him. Finally, moving as unobtrusively as possible, he arrived at the bottom step, hiding behind the bottom stair post. Just then, the cat jumped down from Grandma's lap. As the child Michael watched with mounting anxiety, the cat strolled over to the front door, which was right by the bottom of the stairs, sat down, looked at the door, looked at Michael and . . .

"Meow."

Trapped! Everyone turned to look. Someone said something like, "My, look what the cat found at the bottom of the stairs!"

Smiling, the adult Michael opened his eyes and looked down the stairs, remembering the mixture of excitement and anxiety attached to that event. Most of the details of the event were lost; he couldn't even remember the year. Yet something had occurred to him, something clear and powerful. What were the elements of that memory? Though jumbled, some stood out clearly: the sound of voices; the light from below; his feet in front of him on the stairs; the view from the bottom stair; the cat coming around the corner of the sofa, headed for the door; his fear of the probable consequences. These were not remembered in sequence, really, but only reassembled, like an anthropologist reassembling the scattered remnants of a skeleton. These elements of memory were held together, finally, by a common *time*. Membership in a common epoch, a common era. An epoch and an era indexed by his life. A time that was, though past, still somehow real. A real place, a unique place to which he could travel, by memory. A time with its own unique feel, its own sight, smell, sound, taste. Not the abstract, independent parameter "t" in a secular equation of mathematical physics. A unique, existing, real time.

Michael closed his eyes again and recalled the initial scene – the child Michael sitting there, at the head of the stairs, a safe, dark bedroom behind him, a warm, bright light from the bottom of the stairs below. The reality was palpable, but not in terms of the precision of detailed recall. His memory didn't present him with a photographic image: he couldn't count the toes of the child Michael, or make out the color of his pajamas. Yet, as the adult Michael held the memory before him, its reality grew. It simply *was*. An enveloping, primal sense of place. Also a very personal place. A place in past time in which *he* was. A real, personal child Michael sitting in a real, physical place. In a unique, real time.

The adult Michael opened his eyes and sighed. Speaking of time, it was now time – here and now, in 1992 – to go to bed. It was midnight back in Maryland, and his body had started this day on Eastern Standard Time. Michael reached up and turned out the stairwell light. The stairs in front of him plunged into deep shadow; except for the lighted bedroom behind him, the house was totally dark. Reflected light from the bedroom still marked out the banister, the stairs, the window, the brass light fixture. As his eyes adjusted, Michael could make out his dark reflection in the stairwell window. Colors were less distinct. The green of the carpet was now a dull green-gray. The brass of the light fixture, a dark gray. The white walls were darker too, but seemed somehow to glow, as if reflecting some sort of otherworldly light. The thought occurred to Michael that at very low light intensities a white surface might seem brighter. A white surface absorbs less light than a green or brown one and thus reflects more, but at normal illumination the relative brightness of the white might be less evident. Now, in a much dimmer light, that comparative brightness was more obvious.

. . . a small sound. Nothing, really. A little squeak behind him. Michael's instant thought: *Something-Is-Behind-You.* He jerked around. Nothing. Only the varnished wood floor reflecting light from his first bedroom and the dark doorway of the far bedroom. A second thought: *Something-Is-Coming-Up-The-Stairs.* He spun back around. The stairwell looked back at him, quietly. His skin was alive with a cold, prickly chill. A raw, elemental something reached up, along his legs, up his back, touched the base of his skull. Something fundamental, a dread. Fear of *Something.* The presence hovered, just down stairs. Just under the stairs.

His breathing was rapid and shallow. His mind scrambled for something to hold onto. He spoke aloud, deliberately: "I am Michael Howard. I grew up here but I live in Maryland now with my wife, Carolyn. And her children, Ann and Lynn." The shadow receded. The stairwell was just a stairwell.

He stood up, silently laughing at himself for a minute, then speaking aloud, "OK, Michael. Get a grip. Level out. There is *nothing* here to be afraid of." The sound of his own voice calmed him a bit. He tapped the wall a few times. "See? Nothing there. Just a wall." Laughing more openly, he walked into the bedroom and closed the door. He climbed into bed, pulled the covers in around him, and began to dig into the book he'd brought along: Dante's *Divine Comedy*. Though trained as an engineer, then as a scientist, Michael had been introduced to many classics, including Dante, while on the faculty of St. John's College – a small, unusual school organized around the 'Great Books.' Initially, all he had known about the *Divine Comedy* was that it had something to do with a middle-aged man (Dante) led on a tour of Hell, Purgatory, and Heaven by a Roman poet (Virgil). But once he began to work his way through the complex, metaphysical poetry, the book grew on him.

Packing for the trip to Alton, he had thrown in a prose translation of *Divine Comedy* almost as an afterthought. Now, it served as a good way to redirect and thus drain off the anxiety that had abruptly welled up. He started at the beginning of the book, but soon paused at the line, "its very memory gives a shape to fear!"

> In the middle of the journey of our life I came to myself within a dark wood where the straight way was lost. Ah, how hard a thing it is to tell of that wood, savage and harsh and dense, its very memory gives a shape to fear!

Sitting at the head of the stairs, Michael had been deliberately probing his memory. Were reawakened memories giving renewed shape to old fears? Were the most fearful fears the oldest fears? Fears first shaped from the simple, elemental stuff of a child's world? Fears that were now fantastical to an adult mind? Unreal and therefore unseen and unheeded? And if these old fears were resurrected, made real again, what would their shape be? After a few minutes, Michael resumed *Divine Comedy* where he left off:

> . . . its very memory gives a shape to fear! So bitter is it that death is hardly more. But to give account of the good which I found there I will tell of the other things I noted there.
>
> I cannot rightly tell how I entered there, I was so full of sleep at that moment when I left the true way; but when I had reached the foot of a hill at the end of that valley which had pierced my heart with fear I looked up and saw its shoulders already clothed with the beams of that star – the sun – that leads us straight on every road. Then the fear was quieted a little which had wracked the lake of my heart through all the terrors of that piteous night; and as a swimmer who with laboring breath has escaped from the deep to the shore turns to the perilous waters and

> gazes, so my mind, which was still in flight, turned back again to look at the pass which never yet let any go alive.[2]

Michael leaned back, staring vacantly at the closed window shades in front of him, imagining himself as an exhausted swimmer on the shore: "as a swimmer who with laboring breath has escaped from the deep to the shore turns to the perilous waters and gazes, so my mind, which was still in flight, turned back again to look at the pass which never yet let any go alive." At the head of the stairs, Michael had felt a fear of being engulfed, of being overwhelmed. He had instinctively called his own name aloud – and the names of his wife and stepdaughters – to keep from being swept down. He had reflexively thrown an anchor into common, everyday reality to break his slide down into what? Death? Maybe, though that didn't feel quite right. Whatever it was, fear was a big part of it; "the fear . . . which had wracked the lake of my heart during that piteous night." And yet, it was something to which Michael found himself compulsively attracted, something to which he was powerfully drawn. Again, the poem rang true: When Dante was safely on shore, his first instinct was to turn around and stare at the terrifying place from which he had just escaped, rather than to immediately head toward the morning sun ahead. Or, at the very least, just look forward to the comforting hill while he regained his strength; after all, the only real danger behind him was drowning and he was now safely on shore. So why look back?

Michael continued to the end of the first chapter. Dante, unable to climb the hill by himself, agrees instead to follow Virgil down into Hell, then up the mountain of Purgatory and finally into Paradise: "Then Virgil set out and I came on behind him." Michael paused and returned to the beginning of the chapter, reading it several times until he could reenact in his imagination everything Dante saw and said and did. Like an actor, Michael tried to memorize and thus absorb Dante's experience, to become the character in the story who was about to follow Virgil, the symbol of human reason.

Eventually, midnight caught up with Michael. He put aside *Divine Comedy* and went to sleep.

THE MORNING OF THE SECOND DAY

Saturday, April 4, 1992

MICHAEL OPENED HIS eyes. Even with the door closed and the blinds down, the room was bright with morning sun. The songs of birds resounded in the morning air, echoing among the houses. Michael stretched, swung his feet out of bed, stood up, and stretched again, this time more fully. He opened the shades. The bedroom instantly radiated with dazzling brilliance – sun on the light yellow walls, the yellow quilt bed cover, and the white closet door. Rested and relaxed, he quietly dressed, made the bed, and organized his suitcase, putting the room in order. Bright and now well-ordered, the room seemed peaceful.

He opened the door to the upstairs front room. The dust in the air rendered the incoming beams of sunlight visible. One beam caught the edge of the drawing table, a triangle of light so bright it glowed. A patch of orange enamel paint stood out, a sign of sloppiness in one of his mother's recent art projects – or maybe even his, decades earlier. The orange was so bright it looked as if it were still wet. He could almost smell the rich, pungent odor of fresh paint.

The sounds and the light from outside filled the upstairs rooms, making them seem large, open, and friendly. Michael was wide awake. He took a long, deep breath. It was just so quietly, comfortably *beautiful.* He walked into the front room – his first bedroom – and sat down by the middle of three windows that opened over the roof of the front porch. The glass was old, conveying a slightly rippled image of the front yard, the street and the facing houses. There was a breeze outside. Forsythia bushes were starting to bloom a pale yellow. As the wind brushed the trees, the shadows of the branches swept up and down the sides of houses. Michael was not very good at identifying birds from their calls, but he could tell the distinctive sound of a mourning dove and thought he could also pick out a cardinal. Just then, a bright red cardinal landed on the street light in front of the house and considerately chirped several times.

Yup, that was a cardinal's call, all right.

He tried to open the window, but it wouldn't budge. It was painted shut; so was the one to his left. A large window fan covered the remaining window; that window was still moveable, though with effort. He shoved it open. The sounds from outside were clearer now. The occasional car, the birds, the wind. Also, the distinctive smell of the dust and oil of the fan itself. Michael looked down at the fan, then over to where his bed had been. The drone of this fan had been one of the distinctive sounds of summer. Even now, in Maryland, he would use a window fan as much for the sound as for the breeze – much to Carolyn's amusement. He looked to see if the cord to the fan was plugged in; it wasn't. He thought for a moment about trying out the fan, but then tried to remember where the circuit breaker panel was located. In fact, the house might still have an old fuse box! If so, he'd have to find replacement fuses, as well as the box itself, if the fan motor balked and blew the circuit. He shrugged and closed the window with difficulty. It squeaked and thudded shut. Michael walked to the stairs, the sound of his shoes on the wooden floor echoing in the empty room, then muffled on the carpeted stairs.

In the kitchen, Michael found an open jar of instant coffee in the far cabinet; he was pretty sure it was the same jar he'd bought a month after the funeral when he'd returned to give the house a thorough cleaning. Making a pot of coffee, he downed one cup quickly, then poured himself a second. He walked to the back of the kitchen, switching on the light to the basement. In front of him were the stairs descending to the landing at the back door, where he'd gone out the night before. This time, he continued down past the back door to the basement. The basement had been a big part of Michael's world as a boy. Down here, he had played with railroad trains, model airplanes, and a chemistry set – all gone long ago. What remained was the detritus of a dead house. Everything that had been cleared out upstairs and not immediately junked or hauled back East was here, on one of two tables or on the floor. There were old barrels, mops, brooms, TV trays, baskets, an old fan, and a window air conditioner that had been removed every fall and reinstalled every spring. There was a step ladder on which it would be unwise to step, a box full of old tools, and a rusted saw hanging on a nail. There were two hair dryers: a small one on the table, and a big tall one, the sort you could pull up to a chair and sit under. (Where the hell had *that* come from? Nobody in the family was a cosmetologist. Had Grandpa bought it for Grandma? Michael would never know.) There was an electrical room heater, a broken Toastmaster Oven Broiler, and an incredible assortment of cans and boxes. And there were dozens of glasses and knickknacks. Everyone in the family had taken all the glassware that they wanted. What was left marked the limit where the nuisance of dealing with the stuff outweighed either nostalgia or the suspicion that something just might turn out to be valuable. Surrounding everything were probably a dozen paper bags filled with real, honest-to-goodness trash. Michael hadn't taken care of them when he cleaned out the house shortly after his mother's death because he didn't know when or how garbage was picked up here in Alton, and the people from the bank probably didn't want to take the risk that there was something of value in them. The net result was that the two tables looked as if they were afloat in a sea of trash.

The rest of the basement was less cluttered than the two tables, but still contained a wide assortment of miscellanea: storm windows, screens, chairs. Several towels hung on two ropes that extended the length of the basement. Right in the middle was the furnace, quiet now that it had rid the house of the cold air Michael had brought in with him the night before. When he first arrived in Alton as a five-year-old boy, Michael had been scared to death of that furnace. Not surprising. It was immense. Big limb-like ducts reached out to the rooms upstairs, giving the overall appearance of a metallic gray monster with great outstretched arms. A monster that rumbled. A monster with fire in its belly. Now, with sun streaming in the south window next to the old coal bin, the furnace was just a dusty, fragile set of rusting tubes, duct-taped here and there to keep it intact.

Michael stood on the bottom step, resting his back against the post at the lower right hand foot of the stairs. The table immediately in front of him had been built specifically for his HO model railroad. One October, before that table had been built, Bernie Kohler and he had constructed a haunted house in the then-open corner of the basement. Using construction paper, cardboard boxes, and scraps of industrial felt that Grandpa had brought home from the paper mill, Bernie and he had transformed the space directly beneath him, under the stairs, into a sort of foyer, with spider webs, a chair covered with a white sheet, and a cardboard wall with a hidden panel opening into the hidden rooms of the 'house' of boxes, ending in a graveyard at the far corner. Upon completion of their masterpiece, they invited the other kids in the neighborhood for a tour, with Bernie – always the leader – serving as guide while Michael manipulated the various monsters that would suddenly pop out of a wall or crawl out of a grave. The tours didn't work out like they hoped. The very first kid to take the tour was a good deal bigger than either Michael or Bernie; the 'hidden' panel had to be enlarged several times before she could crawl through. It totally destroyed the foreboding atmosphere they were trying to create. But later that night, after supper, Michael had gone downstairs again by himself. Crawling on hands and knees through the maze of boxes, alone in a dark basement with the feel of cold concrete and the smell of damp cardboard, his imagination quickly began to sense something – just beyond the cardboard wall. Something dark and damp. Something ominous. Something alive. In very short order, the child Michael was back upstairs.

The adult Michael stepped down to the basement floor, simultaneously grabbing the post and swinging himself around till he faced the area under the stairs. He sat down again, this time on the concrete floor, leaning against the back wall under the stairs. This had been the focal point of many of his and Bernie's joint creations – the favored subject being spaceships. Michael had fully imbibed the space-craze of the fifties, which was already full-blown before Sputnik arrived. He read all the usual science fiction authors – Isaac Asimov, Robert Heinlein, Andre Norton – as well as any book he could find on rocket ships. Michael remembered once creating a star-field simulator: a box with a viewport in one side through which he could see the back of the box, painted flat black, peppered with small holes, lit from behind by a small desk lamp. Looking through the viewport, the child Michael had managed to evoke in himself a vertiginous sense of depth, a vision of an immense, awesome star-filled universe just outside his spaceship. He'd get goose bumps. As an adult, he could still remember that sense of awe. In a different way, the haunted house under these stairs had been built to evoke a similar feeling: a sense of mystery, a feeling (creepy, in this case) of something hovering 'out there,' just beyond reach.

Under the surface goal of playing out simple childhood fantasies of Flash Gordon and Frankenstein, there was something deeper: a drive to work out a vague sense deep within the child Michael of a universe full of mystery, something 'out there' to be pursued, to be discovered. Somewhere 'out there' amid the innumerable galaxies were unimaginable wonders. Somehow, he had a place in this mysterious universe.

Over the years, these under-the-staircase spaceships grew more complex. Once, in later years, he had designed a propulsion system based on a nuclear reactor, then drawn a control panel whose dials and switches corresponded to specific functions of that system. He would play the role of flight engineer to Bernie's pilot. (Michael's uncle was a flight engineer with TWA, so it was a natural role for him.) If, say, Michael wanted to create an emergency, he could look at the temperature dial and say, "Bernie! The core temperature is way too high! Push in the control rods! That switch over there, on the left!" – one of a series of switches he had created at the pilot's station. The problem was that the complexity seemed to get in the way of their adventures, rather than help. Michael could still remember his disappointment over that.

Michael's yearnings for a place in an awe-inspiring, extraordinary universe were beginning to run up against the complicated realities of the ordinary world. The endless core temperature dials and control rods.

There were still a few crudely drawn dials painted on the underside of the stairs – though none of them seemed to be a core temperature dial.

The real problem was that reality was slowly becoming identified with practical reality, verifiable reality. Visions of a place in the awe-inspiring, mysterious universe were becoming relegated to the realm of unreality. Flattened into so many Dell Comic Books: *Man in Space, Mission to Mars, Tales from the Crypt, Casper the Friendly Ghost.* Childish silliness. The complexities of adult life demanded your attention; when those complexities entailed pain, they got your exclusive attention. The imperative of doing eclipsed the inspiration of dreaming.

That eclipse cost him dearly. After leaving St. John's College – fuming over some of the faculty there who had told him after two years that he "just wasn't right for the college" – and before finding his current job, there had been a painful period of practical, real-world job hunting. The encouragement of his family had been utterly crucial. He'd certainly been blessed with family support; Ann and Lynn had gleefully offered to burn down the college.

There had been enough money as well. Actually, the most gut-wrenching problem wasn't practical at all – not in the usual sense, anyway. The most gut-wrenching problem was his fundamental lack of direction. During endless student years – going literally back to kindergarten at the old Horace Mann – then as a practicing scientist, then as a college teacher, the deep questions of where he was going with his life were rarely urgent enough to get the time they deserved. Who he was, what his place was in the universe had, to all outward signs, been answered. But inwardly, he had lost his way.

Once, long ago – in this basement, under these stairs – he'd had a sense of the sheer magnificence and majesty of the universe and of his place within it. In later years, he'd see a movie or read a book or hear someone speak, and suddenly the vision would reappear. All at once, he would be staring again right into the star-filled heart of the universe. He'd feel the goose bumps again and know that, despite neglect, the vision was still there. But those glimpses were just that: glimpses of a promise that a right direction for him existed if he would put the work into finding and following it. But he had failed to rework and renew that sense of direction as he matured. There was a widening breach between that powerful but vague sense of his place in a universe of awe and wonder, and all the tangible, practical knowledge he'd piled up in years of adult life. Unattended, the breach had grown so large that he could no longer see where he was. Worse than that, he had completely lost sight of where he wanted to go.

Now, sitting here under the stairs, he began to feel the burden of the task that lay ahead of him. Before long, though, a more mundane burden called attention to itself: his stiff backside. Michael climbed out from under the stairway; concrete walls and floors are hard on middle-aged bones. He looked again at the bags of junk. Should he get rid of them? *No!* He thought to himself that the new owner was getting a good house, full of furniture, for an outrageously low price; the new owner could damn well take out the trash!

...but what would the new owner tell the neighbors about all the junk they'd had to clear out? Feeling instantly guilty, he decided he should take the stuff to a large trash container he'd seen behind the grocery store uptown. With a specific goal in mind, Michael downed the rest of his coffee, bounded upstairs, and grabbed the keys to the car. Outside, it was morning-cool; it promised to be a beautiful early spring day. He moved the rental car to the back door, ferried a dozen bags of trash to the back seat, then returned to the house. His clean-it-up instinct activated and coffee-boosted, he turned on the kitchen radio, finding a Beethoven symphony on KFUO – "Classical 99." He clambered downstairs and puttered around the basement, tidying up this and that, stopping frequently to look at some familiar knickknack. Thus occupied, with the sound of the radio upstairs and sunlight all around, he really felt at home. Finally, it occurred to him that he needed breakfast. He'd go to the Village Inn, after dropping the trash in the grocery store's dumpster, which was within sight of the Inn. He went upstairs to the living room, picked up his coat and began to walk outside when he remembered *Divine Comedy*, sitting upstairs next to the bed; since Carolyn wasn't with him, he'd be eating alone, so he'd probably want a book. A few minutes later, with Dante in his coat pocket, he climbed in the car and drove uptown to the back of the grocery store. As he tossed each of the twelve bags over the edge of the steel trash container, he rehearsed the explanation he would give if anyone asked what he was doing – since he figured what he was doing was probably illegal. Nobody bothered him. Finished, he drove a block and parked in the large lot east of the Village Inn.

The Village Inn had been uptown for maybe twenty-five years; Michael vaguely remembered its opening. One of his Horace Mann classmates was now the owner. Uptown itself was a vestige of the 1800s, when this had been a small college town, distinct from Alton. For a while, it had gone by the informal name of Pie Town; during the civil war, several women had baked pies for the nearby Union troops. Perilous charity in southern Illinois, where the menfolk of some local families were fighting for the Confederacy. Pie Town had effectively merged with Alton in 1911, and became simply 'Uptown'. Michael walked into the restaurant and took one of the two unoccupied stools at the counter. The room was long, with booths and windows lining two sides. The walls were paneled with a deep brown wood veneer. On the wall in front of him, in addition to the milk machine and coffee maker, there were several paintings of chickens, and a flower pot in the shape of a black and white cow. As Michael sat down, the waitress smiled and briskly asked, "Coffee?" He smiled in return, "Absolutely!" Pouring him a cup, she then asked,

"You need a menu?" Michael nodded; she handed him a bright red menu, hard-edged and metal tipped. The banner read, "Village Inn – Open 24 Hours. Breakfast Served Anytime." She returned in a moment and said, "Ready?" He replied, "One egg over easy with sausage, hash browns, and toast."

The waitress was a thin fortyish woman with sharp features. Michael thought to himself that she looked rather old and worn and just a bit like his mother, then suddenly realized he was fortyish too and probably looked a little old and worn himself. This wasn't the first time, returning to his home town after a long absence, that he had been startled to realize that the 'grown-ups' of Alton were now his peers.

The restaurant was packed. A gray-haired man in a conservative brown suit approached the cash register at the end of the counter. The waitress clearly knew him. "How was breakfast, Bill?" she said as she took his check.

"Oh, pretty good I guess. How're things here, Martha?"

"Pretty good." She paused a moment, as if suddenly noticing his suit, then continued, "Well, aren't *you* all prettied up. Coat and tie and all. Looks like you're workin' today." A second short pause, as she counted out his change. "How's the car business?"

"Pretty good. A little slow. Should be picking up here, pretty soon, what with spring and all." He paused. "Isn't that pie case new?"

Michael followed his gaze to a large, stainless steel case behind the counter, half filled with pies. Lemon meringue, apple, cherry, and a dark berry pie – blueberry, maybe, or boysenberry. Momentarily puzzled, Martha looked over her shoulder, then said, "Oh *that*. Bill, that's been here for near two years! It's a good thing we have some *loyal* customers," she kidded him, "or we'd go right out of business." Bill chuckled, pocketed his change and turned to walk out.

Immediately behind him, a lean woman in blue jeans held out a check and said, "You been on board the Alton Belle yet, Martha?"

Martha took the check and opened the register, "Oh, I'll get to that darn boat eventually. Your Ed's been giving me a hard time about it, though."

The woman took her change, and replied, "Yea, he's good for that. See ya." She turned and left.

A moment later, Michael's breakfast arrived. As he ate, he listened to the conversation next to him. A big woman, with graying hair that had once been blond, sat on the end stool of the counter, nursing a cup of coffee and a cigarette.

A few feet from her, two police officers sat in a booth. Their patrol car was parked just outside. One man was tall, thin and quiet, with an expressionless face; he was slowly drinking his coffee. The other, a younger man with thick black hair, was expansively holding forth on recent events; the woman was listening intently.

"Well, we were driving by the high school about six in the morning when I noticed a car sitting way back at the end of the lot. There were two people in the car, so we decided to check it out. They said they were security. I asked the one guy for identification. He wasn't going to give it to me, but the guy next to him said 'When the police ask you for your identification, give it to them!' He did."

"I suppose the school hired those Belleville security guys after that vandalism," said the woman.

"Yea. We try to keep an eye on the place too. In fact…"

He was about to continue when the other officer stood up, saying to his partner, "We need to get going, Jim," then turned, smiling, "We'll see you, Mary."

"See you two tomorrow, Rod," she said, winking to the standing officer. If she couldn't pump Jim for inside information, she could at least reaffirm her insider status with his responsible older partner. Rod smiled more broadly, shaking his head. They gathered their things together, paid their bill and left. Martha refilled Mary's coffee cup.

Michael asked for a piece of cherry pie à la mode. You could almost see the ice cream melting into the pie. It smelled wonderful. As he ate, he listened to fragments of the colorful stories around him: a man working to renovate a red and white '56 Chevy, who carried snapshots with him to show off his progress; a woman upset about the organized crime that was sure to accompany the Alton Belle, the river gambling boat that was now permanently docked downtown; a farmer complaining about the poor county maintenance of Fosterburg road. Michael finished his breakfast and, getting a refill of coffee, continued to listen to the conversations around him. One man in particular told a long tale to Mary about how exterminators drilling holes in his basement to deal with an infestation of termites had hit a furnace oil line, flooding the basement with oil, and forcing him to deal with the EPA since the dirt into which the oil had leaked was now legally contaminated and had to be removed according to federal regulations.

Stories swirled around Michael. Scraps of stories, really. Episodes. Bits and pieces. Random details.

Finally, filled with black coffee and story scraps, Michael asked for his check, paid, and walked outside. The air was brisk and refreshing. He felt good. Wonderful, actually. He had arranged to meet Roy Greywood for an early supper that evening, so he had the day to himself. He crossed the street to an old, stone Presbyterian church with a prominent red wooden door. He had passed by that church countless times as a student on the way to and from Horace Mann, just down the block behind him; in fact, he'd passed it just last night, walking toward the school. The side street paralleling the church dipped and rose to a cross street a block beyond, in front of him. Odd, Michael thought to himself, but he had always walked on the other side, the east side. It felt a bit strange, even now, to be on the west side of the street. Smiling, he crossed to the "right" side, just at the low point of the street, and continued up to the corner. Turning right, he walked up three small concrete steps to the brick sidewalk, perhaps four feet above the level of the brick street. A row of porches lined both sides of the street; the street undulated and disappeared down a hill, ending at a cross street named Seminary. As a boy, not really knowing what a seminary was, he'd thought it was Cemetery, which had always left him with a slightly spooky feeling, as if the residents on that street might include ghosts. Now, as an adult who knew better, he began walking slowly down the uneven brick sidewalk. Trees lined both sides of the street, a breeze from the east moving the branches, a bright network of shadows sweeping the ground and the porches and the roofs. The street was brick as well. The front windows of the houses peered through their porches, shaded from the morning sun.

Halfway down the block, he came to the house of a friend he had known for a while in early adolescence – John Scott. John was now a college professor, somewhere out west. John was a classmate of Michael's friend Bernie and, though Michael couldn't remember exactly how he'd met John, it was certainly through and with Bernie. Looking at the porch and the lawn that descended to a large back yard bordered by a stream – now gone – he remembered vignettes: sitting in the basement reading science fiction magazines, looking through Superman comics, listening to John's Spike Jones records; playing badminton; playing football (poorly) in the large back yard with Bernie, John and sometimes John's dad; playing with the hunting dogs John's dad kept in a simple kennel just behind the house.

One day, Bernie had come up with the bright idea of filling a small plastic ship with homemade gunpowder, then setting the ship ablaze and watching it burn as it floated down the stream in John's backyard. Michael could remember the three of them, working away at the bench in the basement, intoxicated with the power of playing with those chemicals. "OK, now this says we need six parts of saltpeter to one part each of sulfur and charcoal." Just to use the words! Saltpeter. Sulfur. Charcoal. They were entering a new world, they were now insiders. They not only knew about things like saltpeter and sulfur, they were actually *using* them. Wow! And the feeling of danger only heightened the excitement. Bernie had smiled – in his exaggerated way – and said to John, "You know, Bunky, if we mess this up, we're gonna be picking our fingers out of the ceiling." Double Wow! Fortunately, guarded over by the angel that protects curious cats and clueless adolescents, no fingers took any unexpected trips, and after they'd finished stuffing gunpowder into a gray plastic model of the U.S.S. Missouri, they walked gingerly outside and down the hill to the stream. The boy Michael had probably expected a huge, colorful, "KAPOW! KRAACK! KA-BOOM!" – maybe even with the cartoon words hanging in cartoon balloons overhead, like in the comics to which they were all addicted. The reality was an unimpressive hissing followed by acrid, black smoke. The burning plastic smelled terrible. It was a dramatic let-down.

The adult Michael came to the end of the street, which ended at Seminary, and returned to the present long enough to decide which way to go. He crossed the street, headed to the right, and climbed up a hill to one of the stone buildings of what had once been Shurtleff College. His mother had attended Shurtleff for a year back when it was a small private institution; it was now part of Southern Illinois University. He came to a small stone building that had once housed the college book store. He remembered coming here shortly after graduating from eighth grade – figuring he was about to become a high school freshman and thus could *certainly* handle college books. He came across a copy of Virgil's *Aeneid*; there was a Roman soldier on the cover. A war story! Michael loved war stories; he bought it, ran home and began to read:

> I tell about war and the hero who first from Troy's frontier,
> Displaced by destiny, came to the Lavinian shores,
> To Italy – a man much travailed on sea and land
> By the powers above, because of the brooding anger of Juno,
> Suffering much in war until he could found a city
> And march his gods into Latium, when rose the Latin race,
> The royal line of Alba and the high walls of Rome.[3]

He couldn't make heads or tails of it. Over thirty years later, during his brief stint as a college teacher at St. John's, he had to co-lead a seminar on the *Aeneid*; he finally took the old book down from the shelf and read it. By then, the cover was so bleached with age that the title on the spine had entirely disappeared.

Michael continued walking across the lawn of the college, in front of the stone building, to a life-sized, eight-foot, eleven-inch statue of Robert Wadlow – the tallest man in history, according to the *Guinness Book of Records*. Michael's mother had been in Robert Wadlow's high school graduating class. Her high school yearbook had several photos of him, in which *he* always looked more or less normal; it was everybody else who looked strange – *they* all looked like dwarfs! Robert Wadlow died in 1940, at the age of twenty two, of a foot infection due to ill-fitting shoes. Antibiotics were not yet in wide use.

With no place in particular to go, Michael wandered back again to Horace Mann Elementary. Coming to the side door of the school, he was surprised to see it slightly ajar, held open by a small makeshift doorstop. Evidently, something was going on inside. He paused for a moment, and then stepped inside. The stairwell was just as he remembered it. Deep brown and light tan floor tiles; yellow-green wall tiles. He walked up the few stairs to the first floor. The hallway was dark. The sound of muted conversation discernable from upstairs, which accounted for the outer door ajar.

The classroom door to his second-grade room – where he'd seen the janitor the night before – was only a few feet in front of him; he walked over and entered.

The floors. The windows. The dark green 'blackboards' with those horizontal lines. Unchanged. Everything looked worn, old, and hard-used. He looked at the walls, the closets, the desks. He put his hand on one of the desks. The damn things still had ink wells! He stepped back and studied the desk more closely. It certainly *looked* the same . . . Somewhere, deep in his memory, he could see these desks through the eyes of seven-year-old-Michael. The distinction between 'now' and 'then' was gone. Then was now. He said to himself that it was now 1992 – but without conviction. 1992 didn't seem real. It was 1955.

Something had surfaced. A sense of him*self*. In *that* time. Maybe outside time. A powerful, arresting sense of . . . Of what? *Of what??*

Suddenly, from behind him, "Sir? Can I help you?" A man, in his early thirties, stood inside the door, a slightly suspicious note in his voice.

A flood of embarrassment surged through Michael. "Oh, I'm sorry. I'm Michael Howard. I'm just in Alton for the weekend. I went to school here. In fact, I entered second grade right here, when Horace Mann opened in . . . It must have been the fall of '55. I saw the outside door was open, so I figured it wouldn't hurt to peek. Sorry. Sorry."

The man relaxed visibly, and smiled. "Oh, don't worry. I teach fifth grade, upstairs. Several of us are just trying to finish up a project for Monday, and we saw you come in. We weren't expecting anyone, so I just came down to check."

Michael smiled in return, and then pointed to the desk. "I'd swear these are the same desks that were here when I was a student. But how is that possible? That was nearly forty years ago!"

"Well," he snickered, "I don't know. Some of them just *might* be the same. We hold on to things around here forever. Unless something breaks, you keep it." As he turned to leave, he smiled broadly, "Look to your heart's content."

"Thanks," Michael replied.

Alone again in the room, Michael briefly tried to recapture that strong sensation he had felt just a few moments earlier. And failed. Impulsively deciding to leave, he went back out to the street the way he had come in, and then cut around behind the school, walking quickly across the schoolyard toward Seminary. As he tried to recover his earlier thoughts, he found himself questioning his whole 'project' instead. Trying to probe into his past started to feel silly. Stupid. Infantile. Probe for what? In order to do what? He thought about a report that had to be read before the coordinator's meeting at work next week; he had dutifully packed it along with *Divine Comedy*. He could go back to the house right now and probably finish the report by dinner.

He continued to walk. Through the college. Past the statue of Wadlow. He didn't want to abort his project and return to the house to do 'real' work, meaning practical work. Yet he was unable to re-focus on what was, after all, a pretty vague enterprise. Before long, he saw the Amtrak station in the distance and decided to camp out there for a while and straighten out his thoughts. There were no cars in the parking lot. On Saturday, the small station might be empty of Amtrak staff – for a while, anyway. Finally, with a simple, clear goal in mind, the inner turmoil abated. He just walked. Several minutes later, he came to the old brick station house.

The sound of the closing door echoed through the empty waiting room. A small sign at the ticket window said, "Will Return before Noon." He checked the schedule; Amtrak 304, the Abraham Lincoln, was scheduled to pull in from Chicago at 12:10, bound for St. Louis and Kansas City. The station would probably be quiet for a while. Michael sat down facing a window that overlooked the northbound track. As a child, Michael and his mother had taken the train from here to visit his aunt and uncle and their children – his cousins, with whom he absolutely loved to spend time – in Downers Grove near Chicago once or twice a year. Back then, the line had belonged to the GM&O: the Gulf, Mobile and Ohio. GM&O trains were dark red with a thin, gold-edged bright red stripe along the side of the engine and the cars. As a boy, Michael had thought it the height of class. He would stare out the passenger car windows for hours, taking in the soothing rhythms of the train as the flat Illinois country streamed past.

Michael took out his copy of *Divine Comedy* and began with the second chapter. Dante, after eagerly setting out at the end of chapter one, suddenly begins to have second thoughts. He compares himself – unfavorably – to Aeneas (who visited the underworld in Virgil's *Aeneid*) and the Apostle Paul (who wrote in Second Corinthians of his vision of the third heaven), saying that *they* might be worthy of such a journey, but *he* certainly wasn't:

> But I, why should I go there, and who grants it? *I* am not Aeneas; *I* am not Paul. Neither I nor anyone thinks me fit for this, so that if I commit myself to go I fear lest my going be folly. Thou, Virgil, are wise; thou understandest better than I speak.
>
> And as one who unwills what he willed and with new thoughts changes his purpose so that he quite withdraws from what he has begun, such I became on that dark slope; for by thinking of it I brought to naught the enterprise that was so hasty in its beginning.[4]

This scene captured just what Michael had experienced back there in his old classroom, starting off ambitiously to probe his past, then 'unwilling what he had willed.' The motivator of Dante's unwilling was telling as well: both Dante and he had suddenly felt themselves utterly unworthy.

Michael leaned back, looking out the window. Why was he so embarrassed? What was he afraid of? He continued to stare for a few minutes, then stood up and began to pace around the waiting room, speaking aloud to himself.

"OK, now, what was I afraid of back there?" He continued to stare absently out the window, running his eyes along the tracks. "Censure. Ridicule." He reopened *Divine Comedy* for a moment to seek out a passage: 'I fear lest my going be folly.' Folly. Looking silly. Being laughed at. Going off on some damn 'vision quest.' He was nodding his head now. This was on target; the sting he had felt back in that classroom was a projection of censure onto that teacher's initial (and understandable) suspicion. A sudden scene served up by subconscious fears, an apparition of a practical adult laughing at a pathetic quest for childhood memories. How silly.

Michael sat down again and resumed looking out the window. It was pretty clear where some of this fear came from: confrontations with his father. There had certainly been early, defining confrontations that helped establish Michael's reticent, self-doubting character, but those were long since forgotten. Those inarticulate memories lay beneath articulate recall. Michael remembered a much more recent conversation only five years earlier, in which he risked exposing his inner thoughts to his father and received a sound rebuff. But the emotional exchange had unlocked an old injury – and enabled a surprising apology.

Saturday Afternoon

August 29, 1987

Michael grew up with virtually no information about his father. And no curiosity about him either. In his senior year at Alton High, Michael was considering where to go for college. He wanted to become an aerospace engineer and build spaceships; he decided that Purdue University was the best nearby school with a really good reputation in aerospace engineering; Neil Armstrong was a graduate. Then, one morning while he and his mother were eating breakfast, she cleared her throat and said, "Dear, I think there's something you should know." The tone of her voice and the look on her face signaled nervousness. This was costing her. "Dear, your father went to Purdue." Michael stared at her, astonished. He'd had no idea. Of all the schools in the world, he'd unknowingly picked his father's. Incredible!

Still, this didn't prod Michael's curiosity. He just figured it was an odd coincidence, and forgot about it. No big deal. He was seventeen, after all, and unfazed by such things. Twelve years passed. Then, in 1977 Guy called Virginia – completely surprising her – and asked for Michael's address.

A strange communication began between father and son, culminating three years later in a face-to-face meeting, when Michael arranged for a brief stay-over during a business trip. Walking off his plane, he wondered if he would even recognize his father. He saw a tall, thin, vaguely familiar man who smiled at him, walked up, shook his hand and said, "Hello, Mike. I'm your Dad. Good to see you." Back at his father's apartment, Michael met Mabel, his stepmother; he had only learned of her existence a few years earlier, though she had been married to Guy since 1959. Mabel was sitting in one of two large chairs facing a TV; the room was uncomfortably warm. A kindly but subdued woman, Mabel was very solicitous, repeatedly asking Michael if he wanted this or that.

That initial visit had not been unduly awkward at first – no more than would be expected for a father and son who had not seen one another for twenty-seven years. Guy was reasonably pleasant but Michael still found himself involuntarily bracing for an explosion. The explosion never came. What did come was totally unexpected, though it shouldn't have been. Michael and his father were preparing to go for a walk. As he was putting on his shoes, Guy muttered something about doctors and microwaves. Michael leaned forward, asking him to repeat what he had said. He did: "Those other shoes were no good. The doctors turned the microwaves on me until I got a new pair of shoes." For a moment, Michael thought he had simply misunderstood, but hesitated to ask yet again. Then a chill shot through him:

His father thought that the pain of poorly fitting shoes had come from microwaves somehow aimed at him by 'the doctors'! The doctors (from the mental hospital?) had continued to focus microwaves on him till he got new shoes! Later, Guy said that 'the doctors' had put a device in his head that broadcast his thoughts; everyone in the world could hear everything he thought, which is why he hadn't been sent to the front in World War II, since the Japanese would then have known everything he knew about the American military position. Several times over the next two days, when a conversation would veer into this sort of unnerving craziness, Michael clenched his teeth and tried to change the subject. Several times, his father would mutter short sentences that sounded like part of a conversation. Was he hearing voices and responding?

Thankfully, the surprises on this trip were not all distressing. One was the positive relationship that had grown up between Guy and a little girl next door. The day after Michael's arrival, his father took him, Mabel, and the girl – her name was Yesica – to a nearby Dairy Queen, and bought everyone ice cream cones. Yesica's parents were Hispanic, but Yesica's English was flawless. Two of her crayon drawings were attached to the refrigerator, labeled (approximately – she was only six) Grandpa Guy and Grandma Mabel. Guy had pointed them out to Michael with some pride. This was *not* the man Michael remembered. Perhaps there had been some recovery, some move upward.

Still, Michael was immensely relieved to be back on the plane leaving Chicago.

But something was incomplete, unfinished. So, seven years later, near the end of a witheringly hot, humid Midwestern August, Michael returned – several weeks before he began his brief career as a liberal arts college teacher. Little had changed; it was still too hot in the apartment and his father was still crazy. One night, Mabel was preparing supper while he and Guy were sitting in the two chairs in front of the TV. Michael was trying to explain to his father why he was leaving applied science for a teaching slot at an oddball liberal arts college.

"I more or less enjoyed my time as an engineer and then as a scientist. It's fun to absorb yourself in a technical problem. But, finally, my heart wasn't in it. One of the things I wanted from earning my scientist's spurs was to see what it was that experts knew. Science is the premier knowledge-generating engine of our time; I wanted to see what scientists knew and how they knew it. Well, I've gone about as far with that as I want to, or can. I got into philosophy with my non-tech courses at Purdue. I've always wanted to branch out beyond philosophy, to read literature and history. And to find out more about science by actually reading Ptolemy, Copernicus, Newton, and so on. This teaching job at St. John's will give me a wonderful chance to immerse myself in all this and give me experience in teaching as well. Maybe I'm really supposed to be a teacher. Well, I'm going to find out." Michael was still smiling, but paused for a moment; he was beginning to reveal himself but wasn't at all sure he wanted to.

His father was just perceptibly shaking his head, staring at Michael in a strange, unsettling way. When Michael paused, Guy jumped at the opportunity to set his son straight. "Look, son, philosophy is ninety percent bullshit. All that shit Plato and Aristotle said is two thousand years old. It's all obsolete. Those guys all sat around talking about how God did this and God did that. It's all absolute bullshit! God is nothing! Nothing comes from nothing! It's all a perverted lie! We are all animals and nothing more! *Nothing more!* Ministers fill people's minds with sick lies; they make sick brains; they create mental disorders!"

Guy was working himself into a frenzy. His face and neck began to redden. Though still looking at his son, Guy's intense eyes somehow began to shift focus. He was no longer seeing Michael. "And literature! Why would you want to waste your time with stories? Stories are fiction, they aren't truth! What would happen if my mechanic tried to fix my car using a damn storybook? Why do you want to give up science and truth for religion and stories and bullshit?" Guy was visibly shaking. That shaking shook something very deep in Michael. An important moment had arrived.

Michael drew on a hidden reserve of strength, pushed down the deep reaction that had begun to well up, and lunged ahead with his line of thought, "But it's *not* all bullshit!" His father fell silent. Guy's frenzy subsided; his fearsome expression receded. Michael stared at the floor.

A grandfather clock tick-tocked quietly in the background. A decision point had arrived. To answer his father, to expand on what was leading him away from a tolerably successful career path would be to open up something deep inside, something he himself only dimly understood. But to continue with this analysis would also let him respond rationally, let him keep some control over the situation. He looked up. "I think there's something profoundly true in stories. When it really comes down to it, I think we *are* stories. More or less literally. I have no problem, really, believing we are animals as well. I think evolution is pretty well established, so I don't question our being animals; I question our being *just* animals, *only* animals, *nothing but* animals. We are something more, something different, something other than animals, other than material embodiments. Descartes put his finger on it: 'I think, therefore I am.' That we *are*, that we exist as *something*. Utterly simple, but utterly profound. The simple fact: *we are*."

"But when I ask myself what it *is* we are – beyond animals, that is – I can't see much further than the conviction that we are *something*. The best I can do so far is to say that whatever it is, it has strongly story-like features. You said that stories are fiction. OK, I'll accept that; even so-called true stories necessarily select and edit. But good fiction is only fictional at a superficial level. To be sure, Raskolnikov in Dostoyevsky's *Crime and Punishment* didn't really kill the old lady and her younger sister; there *was* no old lady, no sister, and no Raskolnikov. Pure fiction. But when I read that book as an undergraduate, I saw *myself* in Raskolnikov. Raskolnikov was a brooding young student harboring delusions of grandeur.

Hell, so was I! Reading that book was like seeing into myself. At a deep level that story was true. Profoundly true."

Michael paused and then plunged on, not wanting to lose his momentum. Though father and son sat side-by-side, each now stared separately at the floor. Neither wanted to engage the other directly; each, in his own way, had retreated into the safety of his own world. "Let me give you an example. After Raskolnikov kills the old woman and her sister, he realizes the front door of their flat – where he has just murdered both of them with an axe – has been open the entire time. Think of it! You figure you've taken care of all the details. You figure this is going to be the perfect crime. But you don't even think to *close the front door*! Even as you deliberately execute your plan, a part of you makes significant blunders; a part of you *wants* to be punished for what you're doing. You can analyze this psychologically or sociologically all you want; so can I. But to read the story is to experience it, to understand what all this really means. What it means to *be* Raskolnikov. That's why I think that stories are such good vehicles for conveying whatever it is we really *are*."

"Look, Dad, I know this isn't very clear. I only have a tenuous grasp of all this myself, but I *do* have a strong sense that I'm headed in the right direction. I hope I can learn more in this new teaching job, get a little further with this vision of what we really *are*." Michael stopped. He was frowning; he had been staring at the floor for some time without seeing it. Now, suddenly, he was aware of his shoes, the threadbare rug, and his father's shoes next to his.

He looked back up. His father was still shaking his head, but now grinning strangely. "Son, let me say something to you as your father. You're terribly lost. You don't know where you are. You're a bright young man, but you're terribly deluded. You're a *deluded* young man. You and I are both scholars, but I have been a scholar much longer than you. All this bullshit you're getting into is sick. It all comes from mental disorder. Even your use of language is sloppy. It's perverted."

"*Perverted?*"

"Yes, perverted. When you talk, I keep thinking that you aren't using this word and that word correctly. You pervert the words. Your words don't make any sense. What you're saying is all bullshit."

Something inside Michael snapped. He was shaking. Without clearly realizing he was doing so, he stood up. His father suddenly stopped grinning. Michael briefly considered countering his father's charges, but instantly abandoned the idea. Hopeless. Though intelligent, his father had long since succumbed to internal demons and now, wracked with psychological pain, it would be worse than useless to engage him in debate.

Michael abruptly changed the topic. Teeth clenched, he said, "Dad, I'm not a kid anymore, someone you can push around at will." It felt peculiarly like a rehearsed speech he had deliberately come to give. Pausing briefly, he went on, "When I was three or four, you smashed a record right in front of my face. An old 78 rpm record. Capitol Records. The label was purple, with a drawing of the Capitol Dome on it. Do you remember that?" Their eyes met briefly. Then Guy dropped his head, staring down at the floor, drawn and tense. He shook his head, whispering, "No. No."

Guy's precipitous transformation from a frenzied, infuriating adversary to a slumped, broken old man touched something in Michael. He turned and walked to the dining room, where Mabel was quietly laying out dishes. To his amazement, she was nodding vigorously; she whispered, "Now you know what I have to put up with!"

Spring, 1952

A dark, drizzly Sunday. Guy Howard was packing his suitcase. Caustic phrases erupted into his consciousness: "Damn convention"; "Bunch of stupid asses talking about pump design"; "Nothing but fudging coefficients and fitting curves"; "Stupid asses."

Guy was an engineering aide at a large consulting firm outside Chicago. He was still an aide because he had never finished his degree at Purdue. Despite seeing a psychiatrist weekly – an extreme measure in the late thirties – his fragile psyche ultimately buckled under the pressure of a major engineering school. He quit in the middle of his senior year, then found a job with Owens Illinois Glass Company, located in an Illinois river town he'd never heard of before: Alton. Though he started as a draftsman's aide, his intelligence earned him a quick promotion. He met and married Virginia, who worked as a secretary in the main office. Fortunately for Guy, the war came along and gave him a graceful exit from Owens Illinois, where his temper had begun to take its toll. He spent the war in the states as an enlisted man in the army, which sent him to MIT as an electronics instructor. And to several brigs.

His temper again. At the end of the war he found a reasonably decent engineering job west of Chicago, his final army posting. Again, as at Owens Illinois, he initially earned quick praise for his technical ingenuity. And his wife became pregnant with their first child. He began to hope the dark years were behind him.

They weren't. At work, he exaggerated his supervisor's initial praise into expectations of quick success. When this failed to materialize, his swaggering self-confidence collapsed. He became suspicious of everyone around him. At home, his first son died of what was then called crib death – which he promptly blamed on Virginia. Virginia wanted to return to Alton. He angrily refused. He didn't want her near her family, whom he suspected were trying to turn her against him. Michael was born less than a year later.

Guy was plunging into what the twentieth century called paranoid schizophrenia – an objective, clinical phrase that obscures the seething inferno of subjective pain and despair that souls like Guy actually experience. As his control of the external world slipped away, he tried to seize control of his inner world. He reinterpreted, reinvented, and reconstructed that world. Bit by bit, by the simple expedient of willfully reworking the web of private significance of the events around him, he turned failure into victory. An idle comment from Virginia became a blunder exposing her secret, malicious designs – designs his quick mind spotted at once. Bit by bit, he slowly cast himself adrift from the moorings of the real world. As he drifted further and further into a world of his own making, he occasionally experienced a manic sense of power. Of victory over brutal opposition.

But it was a devil's bargain; Guy's willful world was desperately unstable. Willfully distorting the fabric of the interpreted world during fits of anger was adequate while he remained angry, but it's humanly impossible to stay permanently angry. The morning after, it was impossible not to see things for what they really were. An idle comment is really just an idle comment. Common sense and clear thinking are, after all, very robust. To keep to the bargain, Guy had to move deeper into Hell. A place dark and malevolent. And, above all, willful. He taught his intellect to invent temporary cover while his will dug out still darker imagery on which to base his actions.

As Guy was packing, Michael got Virginia to play one of his father's favorite records – Ponchielli's <u>Dance of the Hours</u>, which was already loaded on the turntable. Another father might have taken this as a flattering show of affection from his son, but not Guy. Such a bright picture was ill-suited to the cabal of doleful images from which his will was learning to draw. Strange, how that worked: Dark, ominous, ostensibly powerful images were unable to abide a simple, gentle show of affection. So, rather than turn toward the light, he deliberately dug deeper. Or at least to the limited degree that a fragmented soul can be said to be deliberate.

Consciously he paid more attention to the specifics of packing – looking for a particular gray shirt, a certain dark blue tie – to give his unconscious will time to construct a suitably black vision for action… That name. *Dance of the Hours*. *He was being mocked…*

That was it! Michael had chosen just that record to mock Guy's imminent departure. Guy would leave for the train and Michael would be glad to see his Dad go. In a little while, he would dance. Michael and his mother. They would dance. They would sing. They would laugh at him . . .

Equipped with an enabling vision, he walked calmly and deliberately into the living room, grabbed the record, and smashed it against the table. Michael froze. His mother wavered for an instant, managing her own fear and rage. She looked at Guy, tears in her eyes, "Why, Guy? Why??" Her simple question infuriated him for the simple reason that the demented imagery he had used to motivate his action was incredibly, almost comically fragile. Did he really believe a three-year old had chosen that particular record because of its title? Michael didn't even know it had a title. And would he be glad for his father's departure? Yes, in fact. But again, that simple, devastating question: Why? The answer was really quite simple: Because Dad was unpredictable. Full of manic fun one minute, darkly brutal the next.

Guy conquered the question by evading it. He looked at the clock. 8:30. "Damn it, Ginny, I have to leave in an hour. Where's breakfast?"

Supper was reasonably civil, yet Michael found himself trying to placate his father. After supper, as Mabel cleaned up, he and his father returned to the chairs by the TV. Guy sat down in his chair, picked up the TV remote, and began to talk in a faltering voice. "Sometimes, when I'm changing the channels, Mabel will say something like, 'Slow down! Don't go so fast!' Then the voices tell me to smash the control." He raised the remote, and then brought it down violently toward the arm of the chair, as if he were going to break it. "But I fight them now, so I don't do it. Something like that must have happened with the record I broke in front of you. The voices told me to do it. I'm sorry, son."

Michael stared. His mouth dropped open. "Thank you, Dad! That really helps, that does! That explains a lot! Thank you!!" Michael continued to stand, stunned. A sensation he couldn't describe moved through him. He looked at Guy, who continued to stare at the remote in his right hand. Michael stared at his father for what felt like a long time. Guy continued to gaze abstractly, shaking just slightly. Suddenly it seemed as if something was close; as if Guy was about to cry. He didn't. Nonetheless, something had undergone a subtle shift.

Michael felt like hugging him, but didn't.

Reflecting on this episode later, Michael thought of that broken record as a brutally telling metaphor of his father's soul. The pain and despair of schizophrenia had become so pervasive that even beauty had been perverted. The soothing power of music became yet another threat to Guy's cancerous, demonic inner visions. To keep to the devil's bargain, he willfully broke the record to break its threatening beauty. He willfully broke his own soul.

The next morning, suitcase in hand, Michael walked back into his father's living room for the last time. Mabel was back now in her chair, and Guy in his. Next to each chair was a small bookcase. Guy's contained a wide variety of political, psychological, and sociological books, some of considerable sophistication. Mabel's bookcase was a stark, almost unbelievable contrast, filled with simple books of fundamental Christianity: several Bibles, a Woman's Devotional, and a large pile of evangelical literature. Every morning, Mabel would rise at six for her morning prayers. Sayings of Jesus covered the refrigerator. If Guy had not met Mabel, if Mabel had not stayed with Guy, Michael suspected his father would no longer be alive. At best, he would be a permanent inmate of the nearby mental hospital, all meaningful human contact willfully broken – the 'record' of Guy Howard would have been utterly shattered. Michael smiled down at his stepmother and quietly said, "Thank you, Mabel." Uncomprehending, she smiled back.

Guy began to get up, but seemed to have difficulty; Mabel reached over to help, but Guy angrily waved her off. Arthritis was crippling him, but he refused treatment; he didn't trust doctors. Finally, he got to his feet.

Mabel and Guy accompanied Michael to the door. Guy shook his hand, "I love you, son. Good bye." Michael looked back at him for a second, and then silently put his hand on his father's shoulder. A few minutes later, he drove out the gate of his father's apartment complex. Later, when he first read Dante, he thought of that gate when he read the famous inscription over the gate of Hell in *Divine Comedy:* ABANDON ALL HOPE YE WHO ENTER HERE.

NOON OF THE SECOND DAY

Saturday, April 4, 1992

Michael returned *Divine Comedy* to his pocket, leaned back, and looked out the window. So, what now? It was pretty clear that the free flow of memories had dried up, at least for the time being; he wasn't simply going to float to a renewed vision on a tide of nostalgic feelings. It was going to take work. But what precisely was the task? What was the question he had to answer?

The question that came instantly to mind was the one that had erupted painfully in the wake of his encounter with that teacher: Just what the hell *was* he doing, dredging up memories? What did he intend to do with them? Ouch. This question had the clear potential of leading to an answer he didn't like: It was difficult to see what to do because there was nothing *to* do. Any 'vision' would prove unproductive. This quest really *was* childish and silly.

He took a deep breath, sighed, and began with what seemed like the worst-case scenario. Did he want an impossible fantasy? Did he actually want to go back in time? To return to that classroom, with the inkwells and the green blackboards? Did he want to return to 1955, to be a student again in Mrs. Shea's second grade classroom? Michael let that thought sit for a while, to see if it 'clicked.' There was indeed some attractiveness in it. Mom, Grandma and Grandpa would still be alive. His first bedroom at the top of the stairs would be waiting for him, back at 1918 Washington Avenue. His adult responsibilities would vanish. Yes, there was something moderately attractive in this, but it didn't 'click.' Simply to *be* seven again would be to forego the *memory* of being seven. To lose the perspective that came with that remembering. Somehow, it was the remembering that was important.

OK, so the memories themselves were the real object. But to what end? What was he going to do with them? Was he dredging up these memories just because they felt good? Was all this simply an exercise in nostalgia? This time, he sensed a sort of 'click.' The memories – the good ones, at least – did indeed feel good. Even the bad ones were cathartic – like Mom's death, or that broken record – and thus actually felt good as well.

So, was the objective of all this reflection just the fabrication of good feelings? Michael got up and began again to pace around the waiting room. That still didn't sound quite right; it wasn't simply a matter of feeling good, but assume it was. What then? After an acceptable period of reminiscing, he would return to his job. Only the most Scrooge-like character would deny him the right to spend *some* time reflecting on the formal end of his ties with Alton. The more beneficent conventional wisdom current among his enlightened friends back East would readily grant him the time to 'work through his feelings.' To 'process the grief.' To 'get closure.'

But what *were* these feelings? What did they *signify*? Didn't enlightened contemporary wisdom share with Scrooge the viewpoint that feelings were just things to be managed? Didn't they both view a feeling of grief as a sort of emotional charley-horse? Something to be overcome? Wasn't the only real difference a difference of therapeutic regimen? One solicitous, one Spartan? Didn't they both minimize the significance of the feelings themselves?

Michael was smiling now, shaking his head. This was coming into focus. This was definitely 'clicking.' At bottom, conventional wisdom just supplied us with a user's manual for the human mind: do this to feel good; avoid doing that to avoid feeling bad. But the most important questions weren't *technical*, they were *metaphysical*. The important question wasn't what to *do* with his feelings, or how to *heal* them, it was what those feelings were *about*. What they *signified*. This was the task. The night before, wandering through Upper Alton, he had experienced a strong sense that all these feelings somehow meant something, that they were somehow pointing him somewhere. Something central to who he was, what he should do. Maybe much so-called nostalgia contained this core, this hidden compass. Popular wisdom considered only the withdrawal, and judged those who indulged it as failing to do their duty. AWOL from the adult world. But if properly approached, these nostalgic feelings might hold invaluable insights into who we really are in the world, what our real place is, what we are supposed to do.

Michael's confidence was returning. He walked outside, onto the train platform. After a moment, he stepped onto the rail and carefully started to walk north, balancing for a few feet, then losing his balance and stepping down between the tracks, walking slowly from tie to tie. Then, back up to try balancing again. What were these memories? What did they signify? Where did they point? He looked up, to the spot ahead where the tracks curved out of sight. He closed his eyes and imagined the brute reality of a dark red GM&O diesel sitting on this very track. Its deafening sound, its sharp diesel smell, its awe-inspiring, reigned-in power, about to carry him to his cousins in Downers Grove. From home to home. Along these tracks. His story had been carried on these very tracks. The story of Michael. He walked a bit further along the tracks. The significance, the metaphysical significance of his story. Of any story.

It was time to do some writing. The intellectual questions coming up now were too difficult to pursue with the wandering, free-association technique he'd used to dredge up memories. Perhaps go somewhere and write? With that, he left the train tracks, mind instantly made up, and cut through a small grassy field to Kendall Avenue, a little north of College.

He had been playing with ideas about the importance of the self and of its fundamentally story-like character for years. Maybe this could provide a framework to understand the significance of the memories that had been welling up. For the time being, in fact, he had all the memories he could handle. It was time to begin some serious work, to review his old ideas, and then see if they could provide any keys for understanding these memories, these *feelings*. He could return home and get paper, but that didn't feel right. He'd buy something. A spiral notebook would do it.

Returning to College Avenue, he headed back uptown, and began to play with self-as-story. He had a reasonably clear idea of what a story was. Memory was something like an embodied story, and he was up to his neck in memories. But what about self? What is a self? What am I?

Descartes sprang to mind: I think, therefore I am. Immediately after arriving at this conclusion, Descartes naturally asked himself, "What then am I?" to which his little syllogism suggested an immediate answer: "I am a thinking thing." … the phrase had a leaden feel to it. How about replacing thinking-thing with William James' notion of a stream-of-thought? "I am a stream-of-thought." Taking the stream of thought at face value, some obvious elaborations suggested themselves. As an embodied stream of thought, for example, humans are a part of the physical world, and thus appropriate candidates for a physical self-conception, such as the current scientific view: we are evolved, self-aware products of an immensely vast, immensely old universe. As a member of human society, we can be any of a number of things: a member of a family, a member of a profession, a member of a species: husband, scientist, Homo sapiens. And on and on.

In Michael's view, the problem with all these elaborations on the basic picture of self-as-stream-of-thought was their being *objective*; the most immediate sense of self was *subjective*. 'X is an evolved animal' or 'X is a son, a husband, a scientist' doesn't give much sense of the individual 'X'. But what about, 'X is Michael Howard'? This would mean everything about Michael Howard: his birth in May 1948 in Downers Grove, the trip to Alton in June 1953 with his mother, today's wanderings through Upper Alton. *Everything*. Furthermore, to be intelligible this would have to include the context that provided substance for these experiences. One way to capture all this would be to say that Michael Howard is a story – or, in general, 'I am a story'. What is that? The string of choices immersed in the stream of thought, back through time. A thing-wanted followed by a thing-done followed by another thing-wanted, all against the background of the world, the context of those choices. In other words, a story. Stories have been a powerful vehicle for human expression throughout recorded history and probably long before. It seemed to Michael that the 'Michael Howard-ness' of Michael Howard was tolerably well characterized as a story. Unlike physical, familial, or occupational categories, stories were at least plausible candidates for expressing the actual, subjective *self* in the world.

The problem with this was simple*: It didn't make any damn sense!* What did it *mean* to say "I am a story"?

Michael's opinion on self-as-story vacillated. One day, he'd be entranced by the vividness with which stories seemed to embody something crucially human. Another day, he'd reject the whole idea as sheer double-talk, as the reification of our mental habits. *We* might exist, *we* might be real, but *stories*? What could it possibly mean for a story to be *real*? Wasn't it just a rhetorical flourish on an objective description of human behavior? In fact, wasn't Dad right? When it came right down to it, weren't we just animals? Really clever animals, to be sure, but animals nonetheless?

Michael came to a corner and stopped, waiting for a car to pass. He removed his glasses, rubbed his eyes, replaced the glasses, and exhaled loudly. It was like trying to find your way through a mental fog. But what was he to do? Find a mountain, sit on top, and wait to be struck by inspirational lightning? Not much of a plan. Bit risky too.

Michael suspended his reflections on self-as-story and just walked the remaining three blocks to Kerr's Drug Store, the closest place he could think of that might sell spiral notebooks. Hopefully, writing would put some direction back into his thoughts. Walking in, he looked around but did not immediately see paper supplies of any kind, and asked the man behind the counter. The man looked vaguely familiar... Of course! He was the son of the Kerr that Michael had known as a boy . . . or, no. Maybe even the grandson!

Reaching to a low shelf, this contemporary Kerr – son or grandson – handed Michael a small, red notebook, saying, "This is all we have. I hope you like red." Michael smiled and nodded, paid forty-eight cents, and sat down. He laughed to himself; the notebook had a drawing of an owl perched atop the words, "This Reminds Me." The universe, it seemed, had a sense of humor. Michael sat down at the soda fountain and dug in his coat pocket for a pencil. The small store looked astonishingly like the store he remembered from thirty years earlier. Kerr's sold more than just drugs, to be sure, but it hadn't mutated into a huge, everything-under-the-sun store like most other drug stores. The fountain itself looked like something from another era: ornate soda dispensers, a marble countertop, and a big, old, brass National Cash Register. And Kerr's still had cherry phosphates! Michael promptly ordered one, began guzzling it down, and then paused, to forestall finishing the whole thing in one gulp. He opened the notebook and, between deliberately small sips, began to write:

> Alton, Illinois
> April 4, 1992
>
> Objective: To reestablish a vision of my place in the world.
>
> Working hypothesis:

. . . no, that wasn't right. The word "hypothesis" suggested that this inquiry was going to proceed along scientific lines. The metaphysical quest he had in mind was tougher than that. He erased "hypothesis".

Working 'intuition': What's my place in the world? The answer lies somewhere in the feelings that come up within me here in Upper Alton. These feelings are significant in themselves. They point to something.

Immediate Problem: Reduction of those feelings to psychological states (grief, problems w/father, etc.) or to character flaws (this whole enterprise is just self-indulgence, etc.)

He leaned back, took a deep breath and put the pencil down. Again, that fog. But unless he was going to find a mountain top and meditate, he would just have to continue following up leads and hope he would stumble onto something.

OK. Well, what with all the memories he'd been dredging up, perhaps he ought to think through a specific memory with clearly story-like features. Michael thought back to the episode of that plastic ship Bernie, John, and he had zapped some thirty years ago. What had they been doing? Some obvious answers suggested themselves: internalizing adult scientific knowledge (chemistry, in this case) through play; seeing what it feels like to be a scientist by acting like one; learning how very different reality can be from fantasy – the sickening, acrid smell of real destruction versus the colorful KA-BOOM! of comic book violence. Yet this was all somehow too objective – like the X's in "I am X": X = son, or X = husband, or X = scientist, or X = evolved animal.

Even if improved by saying "X = Michael Howard," what would that really do but make the set of objective things required to specify X very large? The sense Michael was hunting for had to be sufficiently *subjective* to match the very subjective feelings that were coming up in him as he walked around these streets, but even a very large number of objective states didn't seem to add up to a subjective feeling.

He closed the notebook and took another deep breath. He looked down at his cherry phosphate, picked it up and finished it. Then he looked around at the store. There was a poster in the corner for Kodak Film. He remembered a piece of family history about this drug store – in fact, something that even now made Michael feel just a little bit disloyal to be sitting here, in Kerr's. Michael's grandmother loved photography; her photo albums revealed a creative flair and the willingness to experiment. One day, evidently, she had come here to pick up a set of developed photographs. One of the Kerr brothers brought her the photos, but rather than handing them back in the crisp, courteous fashion she'd have preferred, he decided to offer her some off-the-cuff photo criticism. "This is a nice one, Hazel. Good picture of Virginia. Now this one . . . I think you should try to keep things a little more centered. And this one . . ." And on and on. Every picture, according to family legend. The poor man had certainly not intended to be insulting, but what he probably saw as lighthearted banter was seen quite differently by Hazel, Michael's grandmother. That was the last time *she* ever had *her* photos developed at Kerr's!

Michael's grandmother took endless pictures; she had a streak of artist in her, and exuberantly explored the new possibilities of modern cameras – cameras that were becoming cheaper to buy and easier to use. She took one double-exposure photograph of Grandpa pulling himself in a Radio Flyer wagon. Grandma secured her camera in a tripod and snapped one photo of Grandpa sitting in the wagon, then told him to get out, walk to the front of the wagon and hold the handle. A second exposure on the same film now showed Grandpa pulling himself in a child's wagon. Cute.

His expressions in the photograph suggested that he found the enterprise somewhat less entertaining than his wife found it.

Michael thought about the intriguing way in which those photographs combined the objective and the subjective. Objectively, they were the end product of a set of photo-chemical transformations. Every detail within the viewfinder of the camera was represented by a projected image. *Every* detail. Not just the deliberately posed details but the random details that always fill out a real scene: a stained coat, a small stone, a bird just barely visible in the background. Yet, that superabundance, that sheer excess of *objective* detail seemed to capture something of what it was like to subjectively *be* back then, at the precise time the photograph was taken. Looking at old family photos,

Michael often found himself wondering what was going on at the instant the shutter was snapped. What were the people thinking? What were they feeling? The overwhelming wealth of external details recorded by the camera suggested to Michael a comparable wealth of internal details that constitute inner, subjective life. When Grandma had cajoled Grandpa into posing for that double-exposure photograph, his face suggested embarrassment. She had insisted on taking his picture in the front yard, which meant that anyone walking or driving along Washington Avenue would see him. Stiffly holding the handle of an empty child's wagon, or sitting in the wagon, hands clasped around his knees – looking awkwardly off into the distance and wishing Hazel would take the darn photo so he could get back in the house. And those feelings? Those feelings reflected countless earlier episodes in his life, going perhaps all the way back to his boyhood on an Ohio farm. All that background, all that earlier life, depicted in that stiff expression. Michael's grandmother had compiled nearly a dozen photo albums, bringing together thousands of family photographs. Each photo captured the details, the complexity of one instant in time, of one moment in the life of a son, a daughter, a husband, a father, a mother. The countless objective details in the photographs. The countless subjective details in the lives.

Countless subjective details. Countless. Subjective. Details. Michael tapped his pencil on the counter. Countless. (Tap) Subjective. (Tap) Details. (Tap) The significance, the 'something' he was struggling to bring into focus was somehow embodied in the countless subjective details of a life's story. He continued to tap his pencil.

A story was filled with details. That gray plastic boat. The chemicals they had used to make the gunpowder. The table on which they had worked. The comic books piled on the old sofa next to the table. The musty smell of the basement. The light bulb above the table. The daylight coming in the basement window just above the table. As far as the basic story was concerned, though, the boat and the chemicals were all that directly mattered. The basic situation required three teenage boys, three chemicals, one plastic boat, one match, and one nearby stream. That was about it.

Michael's first attempt, a few minutes ago, at explaining the significance of that story had been to relate it to general human patterns. Archetypes, maybe. All children internalize adult knowledge. All children play-act adult roles. All children begin to experience the difference between the simple images of childhood and the more complex images of adulthood. Bernie, John, and Michael were three children. This sort of 'significance' represented them as just particular instances of a general pattern. This was a reductive sort of significance. This was significance modeled on scientific knowledge, seeking the general patterns. Michael smiled to himself. Not too surprising that this would be one of the first ways he'd attack the understanding of significance and attempt to give form to that formless 'something.'

But it was the wrong way. He continued tapping his pencil on the counter. The problem was that significance patterned on scientific knowledge had a habit of doing away with the details. Gunpowder consisted of six measures of saltpeter to one measure each of sulfur and charcoal. Not *this* particular measure of saltpeter, sulfur, and charcoal, but *any* measure. The uniqueness of this *particular* measure was strictly irrelevant. The same limitation applied to sociological/psychological generalizations. *All* children play. Bernie, John, and Michael were just three particular children. Yet the significance Michael was looking for seemed to require a unique connection to these *particular* children.

Michael continued to tap his pencil on the counter. The devil was in the details. The reality of the details. Scientific models did a great job of capturing that reality . . .

No! Scientific models did a great job of *modeling* that reality. What science did *not* do, what it *could not* do was *capture* the reality. The elements of those models were all abstractions. A real scientific theory required a real theorist – a particular, unique scientist – translating the abstractions that constituted the conditions of an experiment into a concrete set of real, unique things; conducting the concrete experiment by following the abstract directions; abstracting the concrete results into the terms of the abstract theory in question. A scientific theory couldn't capture reality. You needed a scientist to do that. A real, live, unique scientist.

For the story of the boat to even have a shot of capturing the reality of Michael's memory, it had to deal with the unique reality of Bernie, David, and Michael. And then the unique reality of everything they touched. Not just any plastic boat, but the very model of the USS Missouri that Bernie, David, and Michael consigned to an unimpressive, hissing, stinking fate. Precisely the boat Michael was remembering, right here, right now.

Countless unique, subjective details. (Tap) Countless. (Tap) *Unique*. (Tap) Subjective. (Tap) Details.

On impulse, Michael glanced at Kerr, who instantly glanced away. He'd been listening to him tap that damn pencil for some time now. Michael grinned. Grandma's revenge! Her grandson was irritating the grandson of the Kerr who'd irritated her.

Michael put away his pencil and paid for the cherry phosphate. "Sorry about the tapping. I was thinking."

Kerr smiled unconvincingly, "No problem."

Michael smiled in response as he turned to leave the drug store. Reaching for the door, he heard the distinctive moan of a diesel locomotive's horn: Amtrak 304 – the Abraham Lincoln, westbound daily from Chicago to St. Louis and

Kansas City – was pulling into Alton Station. Pausing, he looked up at the clock. 12:10 on the dot. Resuming, he tried to open the door – which stuck. In his preoccupied state, he didn't quite know how to handle this practical problem, and paused again. Noting the pause, Kerr called out, "That doorknob's gettin' pretty old. Jiggle it some." Michael tried again, this time with focused attention, and the door opened. He thanked Kerr and walked out into the sunlight.

The door closed behind him. He stood, dazed by the sunlight. Walking toward the car, he kept running through his head that what he was looking for – a vague sense of *something* all around him here in Upper Alton – was a *something* for which no model would serve. That science could not do this was no weakness, really. The job of science was to carve bits and pieces out of reality and find ways to connect them, to model them, and thus to explain them. That it did very well. But today, this afternoon, here in Upper Alton, he had to move beyond that.

It was troubling, but yet somehow exhilarating.

THE AFTERNOON OF THE SECOND DAY

Saturday, April 4, 1992

Michael's car came to a stop by the side of the road. He turned off the engine, sat motionless for a moment with the silence, then opened the door and stepped out. He swung the car door shut. Then, again, silence. Gradually, the faint but persistent sound of a stream emerged. The gravel road cut through a land slowly undulating between the river to the west and flat prairie to the east. A blue jay shrilled, somewhere behind Michael. He began to walk slowly along the edge of the road. The air temperature was in the sixties, maybe even the low seventies; the land was dry and warm. There was something transfixing about this quiet, simple landscape. Following the sound of running water, he came to a modest stream winding through a grassy field next to the road. On the near shore stood a gnarled old black oak, the buds of its summer leaves just beginning to swell. The tree looked as if it were shrouded in a pale green mist.

Brooks and streams in farm country always filled Michael with a sense of peace, a sense of security, a sense that he belonged in the world. As a teenager, he'd ridden his bicycle out here for hours. He loved the look of things.

The way a tree looked, at the side of a road. The way a road looked, climbing a hill, disappearing over the crest. The way the crest of a hill looked, against the sky. It always felt like there was something profound just beyond that hill. Just where the horizon met the road. Decades later, he would occasionally play hooky from work and drive out into the western Maryland countryside. Though hillier, it was still reminiscent of the Illinois countryside. So today it was only natural that Michael would decide to drive out to the country to think. To try to sort out that insight about how 'countless unique subjective details' might help him get a clearer sense of what these feelings were trying to tell him.

Michael walked over to the oak; looked for a long time at the water rippling over small rocks, running alongside the grass. Exposed by years of channeled rain, the roots of the tree against which Michael leaned had been dried to a dusty, grayish brown by the sun, down to within a few millimeters of the water's surface, where occasional waves kept the roots damp and black. He found a natural seat formed by the roots of the tree and sat down, resting one foot on a rock in the brook, the other on an opportune tree root. He sat down, leaned against the trunk. Tossed a few twigs into the water. Watched them maneuver their way around the rocks. He focused on a small standing wave, immediately downstream of a particular submerged rock.

The shape of that wave depended upon only a few features of the rock, along with the speed and depth of the impinging stream. The simplicity of science. That rock could have been one of a million slightly different shapes and still generate a standing wave – a standing wave even more nearly the shape of the real wave than the slightly-different rock was nearly the shape of the real rock. Nature often averages things out. And time? The wave was a standing wave. It wasn't moving with respect to the shore – or with respect to Michael, for that matter. A scientist could take the basic equations bequeathed by Newton, make a simple coordinate transformation, and eliminate time altogether. The predicted, time-independent wave shape would still be a very good match with that wave there, in that stream. Not an exact match, just a very good one. Unlike the more faithful reproduction of that wave one would get in a photograph, a theory of science only gave you the pencil figure sketched onto the photograph, along with a coordinate system and a few well-defined variables. Scientists don't take photographs, they make drawings.

Michael sighed and leaned more heavily against the tree. Today wasn't a day for science. On a branch, above him, a squirrel fussed at some affront. For a small animal, a squirrel could sure raise a racket. He shook his head and leaned back against the tree. He deliberately stopped thinking about anything, and just looked at the stream in front of him.

Sunlight rippled hypnotically on the water. Michael closed his eyes and listened to the stream. He opened his eyes several times, then settled into a state not quite awake, but not quite asleep either. As he opened his eyes once more, he noticed that the squirrel had resumed its chattering, and wondered about the cause. Across the brook, blue jays were diving on something in the grass just beyond the wooden fence that bordered the opposite bank. Two of them would alternately dive toward the grass, then veer back up and return to an overhanging branch of the oak. A big, gray tabby cat trotted out from under the fence, his ears back, his dignity wounded. This was obviously what had the jays upset.

"Psst! Hey pussycat! Over here, buddy!" The cat looked across the water and stared for a moment. Then his ears relaxed, his tail went up, and he deftly crossed the brook. Obviously, somebody's cat; very used to people. Very familiar with the landscape too, judging from the way he'd crossed that brook. The cat trotted up, meowing. Michael held out one finger. The cat took a few trial sniffs, then rubbed his head along Michael's hand, around walked behind him – rubbing all the way – and finally lay down, stretching out in a posture that clearly said, "Pet me!"

Michael petted the cat, who purred more loudly with each stroke. "Hey, Tomcat, how's it going? May I call you Tom?" Michael checked briefly to see if Teresa might be a better name; it wasn't. "I'll just go ahead and call you Tom, if you don't mind. Are you out on your rounds, Tom? You're clearly used to us humans, and I'd say that whoever owns you is pretty good to you. We have three cats back home in Maryland. Gizmo, Wizard, and Bubba. You look and act sort of like Giz." The cat purred and looked up at him; Michael just watched him for a while, and then began to wonder aloud, "Tom, what are you thinking? When you look at me, what do you see?"

The cat looked up and continued to purr. Then as if suddenly remembering, "I'm late! I'm late! For a very important date!" he looked around, back across the brook. He got up, trotted a few yards, and stopped to sharpen his claws on a root. He looked briefly at Michael, then again across the brook, and hurried off, re-crossing where he'd crossed a few minutes earlier. Then, just as suddenly, he stopped to scratch and clean his backside – maybe a flea was objecting to the abrupt ride. Cleaning finished, he trotted off along the far edge of the brook, stopping to sniff things here and there, and eventually disappeared into the tall grass.

Michael resumed thinking about the cat – about Tom. Did the cat have a self? Was there an existing "I am" in him? The cat certainly knew things about the world, felt things about the world, acted deliberately and purposefully in the world. How close were the Tom-self and the Michael-self? Michael remembered an episode in high school in which he had wondered about something similar. Sitting upstairs in his second bedroom one warm September night, Michael's cat – Sam – had jumped onto the desk and stood right in the middle of Michael's open high school geometry book. Capitalizing on the opportunity to procrastinate, Michael had begun a one-sided conversation: "Well, Sam, if you're going to stand on that chapter, then you'll just have to tell me about congruent triangles!"

Sam was unimpressed.

This had raised an intriguing question: Though Sam was a smart cat, he would never be able to understand anything about congruent triangles. On one hand, of course, this was obvious. But as Michael began to think about some of the simpler theorems of geometry, they were so obvious that he couldn't understand how they *couldn't* be understood. They were so . . . *obvious*! Two triangles, each congruent to a third triangle, were congruent to each another. Of course! They were exactly the same size and shape as a third, so they *had* to be equal to one another. So, how could Sam not know this? Michael hit on the idea of two mice running one after the other into the same mouse hole, each just barely squeezing through. Surely, Sam could figure out that both mice were the same size. Wow! That did it!

. . . until Michael began to think about it. He could imagine – as a cat – wanting to chase a mouse. He could imagine wanting to kill one. He could even imagine (with difficulty) wanting to eat a mouse. But try as he might, he couldn't cat-imagine the abstract thought "same size." Sam's understanding and Michael's, though both adequately represented three-dimensional space, were forever different. There were truths about three-dimensional space that Michael understood in a way Sam could never grasp. He thought about that for a moment.

Then it hit him. If there were ideas that Sam could never, ever understand, were there ideas *he* could never, ever understand? That *no* human could ever understand? Once the possibility formed clearly in his mind, it seemed irrefutable. As obvious in its own way as the theorems of geometry. Millions of years ago, when Sam's prehistoric ancestors were trotting around looking for prehistoric mice, all the theorems of geometry were still true. There was simply nobody around yet with the mental hardware to comprehend them. So, what was there to prevent the existence of super-theorems in some incomprehensible (to humans) super-geometry? True but utterly beyond human understanding. Michael looked out the window into the night and thought about that for a while – feeling a little uneasy about it.

The same year he was talking to Sam and wondering about the unknowable, his geometry teacher loaned him a book called *Flatland*. Written by an Englishman in the nineteenth century, the book narrated the adventures of a resident of "Flatland," a place with only two dimensions. A Flatlander could move forward and backward, or to the right and left, but not up and down. The book had a lot of fun with the idea of how circles and squares and angles would look to a Flatlander. But the most gripping part of the story came when the Flatlander – call him T. D. (Two Dimensional) Sam – was suddenly transported up into Spaceland. There he saw triangles from above. He was shocked, but eventually able to take in some of what he could now see. When he returned to Flatland, though, he tried to tell others about the mysterious third dimension and got thrown into jail for spreading seditious rumors. It turned out that the idea of a third dimension was considered to be dangerously subversive by Flatland's leaders.

But why was poor old T. D. seen as subversive rather than crazy? 'Up' doesn't exist in two dimensions, so why weren't T. D. Sam's claims simply unintelligible? *Because something in his claims 'rang true' to the Flatlanders*. They couldn't express it very clearly to one another, but in a vague, intuitive way, it 'made sense.' What's more, this vague 'something' was communicable, it was (amazingly!) expressible in words. If T. D. had come back and said, "Hey, everybody, triangles are really jnmrtgb! They actually have lpwtrffcx!" he'd have been given a sedative, not jailed. Instead, he might have said something like this: "Look, chaps, you know what it means to go forward and backward along a line. Right? Now, imagine standing at a certain point on a line and stepping just the tiniest bit left or right, off the line. At a right angle to the line. You haven't moved forward or backward one whit, yet you've certainly moved! You've moved in the other dimension, the second dimension! So, just imagine you are standing at a point, not moving forward or backward, left or right, but . . . but moving some other way! Call it a 'third dimension.' I know you can't understand directly what a 'third dimension' is – neither can I, really – but certainly you can see by analogy what that way must correspond to!" A sympathetic jailer, well-schooled in (two dimensional) geometry, might scratch his chin and admit, "I find this to be an oddly compelling argument!" – after looking over his shoulder to be sure nobody else was listening. There might be something beyond two dimensions, even if we could not really ever know it. Know it with any clarity, at least. But it still might exist.

Michael sat by the brook, thinking again about the unknowable, resuming a thirty-year old train of thought. It was surely irrational to claim that everything in existence was potentially knowable to us, explicable by us. The unknowable existed; the inexplicable existed. It might even be apparent, in some vague way. Something elusive that still 'rang true.' The idea that unknowable truths might exist was not too hard to grasp, once the possibility was raised.

He leaned back against the tree. Thinking. Looking at the tall grass beyond the fence on the far side of the stream, into which Tom had so recently disappeared. Cats had been prancing through tall grass for millions of years, much longer than Homo sapiens had been around to prance through anything.

Go further back. Ice age mammals; Jurassic dinosaurs; Cambrian trilobites. The Precambrian emergence of life itself. We Homo sapiens are really very new. On the timescale of the universe all of our science and all of our knowledge are really only the first babblings of a newborn. And that's not just a rhetorical flourish. It's simple truth. The universe is very big and very old, and we are very, very young. Strict, sober truth.

A simple inference from that truth: the scope of reality must vastly exceed the reach of our knowledge. Yet this inference does not upset or undercut our knowledge; indeed, it stands upon that knowledge – knowledge that the universe is very large and very old.

There must be, then, a vast gap between knowledge and reality. The vague something that Michael sought might well lie in that gap. The sense of having a place in the universe – of having a significance in the universe – was simply a sensing of something in that gap. Perhaps one ought to say that there was a significant gap between knowledge and reality. The understanding of self – whether as story or as countless, unique, subjective details – might depend crucially on realities beyond the reach of reason, realities beyond the gap. Beyond knowing but nonetheless real.

It began to dawn on Michael that he had just found the key he was looking for.

Funny, how that worked. Sometimes when he had been thinking hard about a problem, the solution would present itself more or less fully-formed, as if handed to him. And it always seemed to arrive in two stages. First, he would see the answer. Almost *recall* the answer. Without emotion. As if he were just remembering an old, familiar idea. Nothing to get worked up about. He'd known it all along, after all. The answer. Right...

Then, there would be a surge of emotion.
. . . *Right! The answer!*

Our sense of having a significance in the universe is simply a sensing of something in the gap between knowledge and reality. For us, then, a very significant gap. The sense of significance is precisely the sense of my feelings pointing beyond themselves into something unknowable, something that transcends ordinary knowledge. In other words, something transcendent.

Euphoria swept over Michael. He deliberately tried to focus it into words. He hastily pulled out his notebook:

> Our sense of having a place, a significance in the universe is simply a sensing of something in the gap between the reach of knowledge and the scope of reality. For us, a very significant gap. The sense of significance is precisely the sense of my feelings pointing beyond themselves into something unknowable, something transcendent!

Michael read what he had just written. Perhaps it would be better to say the reach of reason, rather than the reach of knowledge. It's our reason that does the reaching, not our knowledge; he erased knowledge and replaced it with reason: Our sense of having a place, a significance in the universe is simply a sensing of something in the gap between the reach of reason and the scope of reality. Ideas were flowing quickly now; having amended his claim, he immediately saw an implication.

The first thing that strikes me is that this gap is blindingly obvious. Once seen, it is pretty much impossible to dispute. The big catch is that when I know something specific – that I see a cat in front of me, for instance – the gap is nowhere in evidence. I don't cross my epistemological fingers and add, under my breath, that my knowledge is provisional, that reality might have conspired with my eyes to give me the illusion of a cat. I know there is a cat and I know there is a gap; the two are complementary. Like orthogonal (i.e. perpendicular) vectors in a plane, they each carry valid but independent information.

The gap is blindingly obvious only if I look for it. It's as if I were inside a house, looking out at the universe through a glass window. Unless I consciously thought about the window, I wouldn't see it; I'd only see what was outside. If I saw a cat outside, I'd never think about the glass. I'd just say – quite correctly – "I saw a cat. Reminds me of Gizmo." I would know what I saw, without qualification. Shifting my focus, I would see the window – at the expense of seeing Gizmo with any clarity. The knowing of one and the knowing of the other are complementary, not conflicting.

Michael was nodding his head now. Yes, this all made sense. The gap was obvious, but easy to miss. The inherent limitation of knowledge implied by the gap did not undercut knowledge itself: it complemented it.

He leaned back against the tree and stared for a while at the brook. It began to feel as if his work here was done. That to stay here longer was not right. Almost, in an odd way, sacrilegious. He stood up. Looking down at the water flowing among the rocks at his feet, something had undergone a subtle shift. No particular fact seemed to be different, yet somehow *everything* was different.

The top of the mountain was apparently at the bottom of a brook.

He looked over at his parked car, and decided to go to Hayner Library to do some more writing. Maybe look at a book or two. He drove back to Alton, soon coming to State Street. Starting in what was known as North Alton, State was one of those streets that began on the level of the surrounding Illinois countryside, then descended sharply to Downtown, along the riverfront. During serious floods, the last block of State was usually underwater. State Street's sharp descent began at the St. Peter and Paul Catholic Church. The river ahead was just huge! If years away from home had made the house at 1918 Washington look smaller, it had made that river look bigger.

Jenny D. Hayner Library was just three blocks short of the river-level terminus of State. The library's main branch, Downtown, was now housed in two buildings. During high school, Michael had worked evenings and weekends in the older building – at that time the only building. Built in Old English Gothic style, with a turret overlooking the corner of State Street and Fourth Street, its imposing vaulted oak and pinewood ceiling gave the library a church-like feel. Michael parked on State Street, outside the old building, which now housed children's books and the library administration. The newer, adult library was a block east, down Fourth Street.

Michael walked to the renovated front door. It wasn't quite as forbidding as the one he remembered. Inside, the place didn't look terribly different. Now dedicated solely to children's books, the layout was basically similar to that of thirty years earlier; there were stacks in the back and a main desk in the cavernous central room. He watched a gangly young boy in a garish T-shirt check out a huge pile of books, angrily refuse assistance from what looked to be an older brother, then stagger past Michael to the door. Michael hid his smile from the boy; he had the right to fight his boyish battles without adult condescension.

Once the two boys had gone, Michael walked slowly into the library. A beautiful place, filled with books written by adults and read by children. How many of those adult authors were secretly trying to reach back to their own childhood, trying to see the world through their own child's eyes? Yet not wanting to *become* children, not wanting to *forego* the educated perspective of adulthood but somehow to meld the two. To remember their lives in perspective, in context – in much the same way Michael was trying to remember his.

He continued through the enormous central room to the stacks, looking for the science fiction section. He found it next to windows facing Fourth Street. As he expected, a number of old books were still there. After seeing ink wells at Horace Mann, he figured the library might just follow the same 'if-it's-still-in-one-piece-keep-it' philosophy. He looked for something by Andre Norton, and found *Star Rangers*. Its prologue was on page three.

PROLOGUE

There is an old legend concerning a Roman Emperor, who, to show his power, singled out the Tribune of a loyal legion and commanded that he march his men across Asia to the end of the world. And so a thousand men vanished into the hinterland of the largest continent, to be swallowed up forever. On some unknown battlefield the last handful of survivors must have formed a square which was overwhelmed by a barbarian charge. And their eagle may have stood lonely and tarnished in a horsehide tent for a generation thereafter. But it may be guessed, by those who knew the pride of the men in their corps and tradition, that they did march east as long as one still remained on his feet.

In 8054 AD history repeated itself – as it always does. The First Galactic Empire was breaking up. Dictators, Emperors, Consolidators wrested the rulership of their own or kindred solar systems from Central Control. Space pirates raised flags and recruited fleets to gorge on spoil plundered from this wreckage. It was a time in which only the ruthless could flourish.

Here and there people tried vainly to dam the flood of disaster and disunion. And, notable among these last-ditch fighters who refused to throw aside their belief in the impartial rule of Central Control were the remnants of the Stellar Patrol, a law enforcement body whose authority had existed unchallenged for almost a thousand years. Perhaps it was because there was no longer any security to be found outside their own ranks that they clung the closer to what seemed in the new age to be an outworn code of ethics and morals. And their stubborn loyalty to a vanished ideal was both exasperating and pitiful to the new rulers.

Jorcam Dester, the last Control Agent of Deneb, who was nursing certain ambitions of his own, solved in the Roman manner the problem of ridding his sector of the Patrol. He summoned the half dozen officers still commanding navigable ships and ordered them – under the seal of the Control – out into space, to locate (as he said) and re-map forgotten galactic border systems no one had visited in at least four generations. He offered a vague promise to establish new bases from which the Patrol might rise again, invigorated and revived, to fight for the Control ideals. And, faithful to their very ancient trust,

> they upped-ship on this mission, shorthanded, poorly supplied, without real hope, but determined to carry out orders to the last.
>
> One of these ships was the Vegan Scout – *Starfire*.[5]

This was classic, adolescent space opera. It was like *Star Wars* with its ominous empire and its ragtag Jedi Knights, incarnate here as the Stellar Patrol. Michael could still remember the opening scene, in which the Vegan Scout *Starfire* crash-landed on an unknown planet. In the course of their many adventures, the crew discovered signs of an ancient, vanished civilization. One curious fact surfaced early in the story: the green of the vegetation and the blue of the sky were surprisingly beautiful. Strangely familiar. Then, at the end of the book, they discovered that this planet was the forgotten home of the human race – *Earth!*

This was the real stuff, all right. A vision on a galactic scale. Even more, a vision placed confidently in a majestic sweep of time. The plot device of making an unknown planet in a "forgotten border system" into Earth six millennia in the future could really grab the imagination of a young teenager. It had really grabbed Michael's. He and Bernie had probably acted out some version of the story from the cockpit of their under-the-stairway starship. However, beneath the adolescent veneer of *Star Rangers* was a credible vision of humanity. A seeing of life on a cosmic scale. Michael studied the book cover's colorful, surreal landscape. Very dramatic.

Very typical of science fiction cover art of that era. He tried to bring back what he had felt, looking at perhaps this very cover more than three decades earlier.

The memory surfaced quickly: a powerful sense of a strange, far-off, alien land. And of something a little frightening, a little scary. Yet something that was intensely, almost painfully compelling. And of *time*. *That* was crucial. The image wasn't compelling just because it was exotic and alien. In fact, that wasn't even a major part of it. The image was compelling because it encompassed the birth and death of whole civilizations. It presented the adolescent Michael with a vision of epic time, of uncountable eons, of infinity.

He thumbed to the end of the book. This was the point where the main character, Sergeant Kartr, realizes that this borderland planet was the home of the human race. His home as well: "Terra of Sol – humanity's beginning!" He comes to the ruins of an ancient building, the Hall of Leave-Taking where humans from Earth had originally departed to explore the galaxy. The book closes with the lines, "Kartr turned his back upon the Hall of Leave-Taking and ran lightly down the eroded steps. The wind was chill but the sun was warm. Dust puffed up from beneath the marching feet. "Yes, the end is not yet! Let us go!"

Michael winced a bit. Language well-adapted to an adolescent's ear was a bit overwrought to an adult's. Nonetheless, the underlying image was as powerful now as then. He put *Star Rangers* back, and looked around for some other old book. Just a little further down the shelf was *Mists of Dawn*. By Chad Oliver. Hmmm… Pulling it out, he instantly remembered its cover: a man stepping out of a spherical time machine onto a winter landscape with three Neanderthals looking on in amazement. He'd forgotten the title and author, but that *picture* immediately brought back the memory of sitting in his living room at 1918 Washington, mesmerized by the vision of a landscape fifty thousand years old. A time when even the pyramids were in the far future! He couldn't remember much of the story, and thumbing through the book didn't bring anything back – there was nothing quite as clear as that Prologue in *Star Rangers* – but the overpowering sense of vast sweeps of time was still clearly etched in his memory, and attached to that cover picture.

There was something about such visions which fit with transcendence, with something beyond the knowable. The sort of visions created by Andre Norton and Chad Oliver were clear only to a point. The distant future and the distant past were both beyond conventional knowing. Thus, their contemplation more or less automatically seemed to invoke confronting something beyond us. The placing of an event in time. *Deep* time.

He thought about that. As an adolescent, one of the ways we recognize our place in deep time is by means of epic sagas. They touch the part of us that recognizes itself as part of something larger. But as we age, another method opens up for the exercise of that faculty of recognition: memory. Perhaps the common feature in adolescent science fiction and adult memory is that both place us solidly in time, stretched out between past and future.

That second, mature method – adult memory – has a distinct advantage: it employs the actual, unique elements of a life. Not just life in the abstract, not just the sketch of a life in a novel, but our unique life, the one we have actually lived. Yet there is still, in that earlier adolescent method, something crucial, something flowing into our lives from outside, from the universe. The melding of these modes provides one of the more profound satisfactions of maturity.

Michael re-shelved Oliver's *Mists of Dawn*, then turned around to look out the window to Fourth Street, perhaps twelve feet below. And smiled, remembering a little adventure.

He and another library aide, Matt, had been shelving books on a warm spring day not unlike this, over twenty-five years ago. The window – that window, right there – had been open. A big dump truck had rumbled up Fourth Street, slowing to a stop at State Street. Matt had just picked up *A Choice, Not an Echo*, by Phyllis Schlafly, a conservative author who lived in Alton. Matt hated Schlafly and, with the untroubled assurance of high school convictions, decided to do something about it. He leaned out the window and tossed the book into the dump truck. Michael stared at Matt, who nonchalantly said, "I'm just checking it out. Long-term!" On the whole, Michael had been a pretty restrained teenager. Matt was much quicker to push the edge. Though uneasy with Matt's impulsive censorship, Michael had to admit he was impressed by its boldness.

Still smiling, Michael went back through the central room, left the old library, walked past the turret at the corner of State Street and Fourth, and continued on down the block to the new building – a renovated grocery store. This building was level with the terminus of State Street. River level. Inside, Michael picked out a table and hung his coat over the back of the chair. The new library was modern and neat. Bright, too. Large windows along two sides looked out onto Fourth Street and Belle. On impulse, he walked to the reference desk, staffed by a young librarian and a girl he took to be a high school library aide. Michael asked the librarian if she knew whatever became of Bea Wright. She smiled, said "No" politely, and looked at Michael with a quizzical 'why-should-I-know-Bea-Wright?' expression. Oops. He tried to explain that Bea Wright had been the main librarian back in the early sixties. This young librarian's eyes widened just a bit as she said "No, I haven't met her," and smiled again, with a smile Michael interpreted as 'I-haven't-met-Abraham-Lincoln-either.' Michael retreated to the stacks. The nonfiction Dewey Decimal 100s were the closest.

Michael had never really gotten used to Library of Congress filing – the filing system typical at college and university libraries. One nice thing about the Dewey Decimal 100s was that philosophy lived here, a short reach upstream from the bloated lowest numbers – everything you ever wanted to know about the zillions of computer-related products that did not yet exist when Michael had been charged with the care and feeding of Hayner's books.

He found a copy of Rene Descartes' *Meditations*. Dewey decimal 194 DES. The book was worn and evidently near the end of its serviceable life. *The Meditations* were a natural for Michael this afternoon. In them, Descartes had focused with laser-like intensity on questions of knowledge and the knowable. "What can we know?" More to the point, "What can we know with *certainty*?" *The Meditations* had helped swing philosophy into its modern epistemological course. Michael slipped the book under his arm.

Sharing the same shelf with Descartes was Blaise Pascal. His famous *Pensées* lived at Dewey decimal 194 PAS. A countryman and near-contemporary of Descartes, Pascal was a scientist first and a philosopher second. Michael had first encountered the scientist; while at Purdue, Michael routinely worked with the modern metric unit for pressure – Pa, the Pascal. (One Pa equals three ounces per square yard, if you're interested.) Decades later, upon joining the faculty at St. John's, Michael was quite surprised to learn of Pascal's philosophical side.

Pascal's intellectual gifts were evident early in life. There is a story that one day at the dinner table, young Pascal noticed that tapping a dinner plate made an exceptionally clear, ringing sound. He was fascinated, and proceeded to walk around the house, tapping everything in sight. He disappeared for a several hours and returned later with a little essay on vibration and sound.

Pascal's father, Étienne, was in charge of his son's education. Étienne exposed his son to all the major areas of learning except one – one he thought too intoxicating for a young boy. Mathematics! Étienne carefully hid all his math books in a locked closet so Blaise would not be tempted. One day, though, Étienne discovered Blaise out back of the house furtively reading a geometry book. According to family legend, Blaise was busily working out many of the theorems of geometry for himself.

Then came the night of November 23, 1654. Pascal's biographers refer to it as "The Night of Fire," because a vision of fire figured prominently in his mystical experience. His life was utterly transformed; he became one of Christianity's most articulate spokesmen. Yet, he evidently *never mentioned* this mystical experience to anyone during his life! It was only after his death that a servant discovered a record of the events of that night sewn into the lining of his coat. Ever the scientist, Pascal had carefully recorded his experience, as best he could. Later, he copied his hasty notes onto the parchment which was found in his coat.

After this experience, Pascal determined to write a book that would present and defend the Christian faith. He didn't live to complete his project, but after his early death at the age of thirty-nine, his unfinished notes and essays were brought together in a book called *Pensées* (a rough English translation would be *Thoughts*), published some three decades after *The Meditations*. Even in its unfinished state, *Pensées* became a classic. Among its wide-ranging topics were several famous quotes about the limits of knowledge and of reason. Another natural for Michael's afternoon reading. It joined Descartes, under Michael's arm.

These two were enough for now. Michael returned to his table, opened The Meditations, and read its famous opening lines:

MEDITATIONS ON FIRST PHILOSOPHY

in which are demonstrated the existence of God and the distinction between the human soul and the body

FIRST MEDITATION

What can be called into doubt

> Some years ago I was struck by the large number of falsehoods that I had accepted as true in my childhood, and by the highly doubtful nature of the whole edifice that I had subsequently based on them. I realized that it was necessary, once in the course of life, to demolish everything completely and start

> again right from the foundations if I wanted to establish anything at all in the sciences that was stable and likely to last. But the task looked an enormous one, and I began to wait until I should reach a mature enough age to ensure that no subsequent time of life would be more suitable for tackling such inquiries. This led me to put the project off for so long that I would now be to blame if by pondering over it any further I wasted the time still left for carrying it out. So, today I have expressly rid my mind of all worries and arranged for myself a clear stretch of free time. I am here quite alone, and at last I will devote myself sincerely and without reservation to the general demolition of my opinions.[6]

Michael took out his notebook. Flipped back to the first page:

> Objective: To reestablish a vision of my place in the world.
>
> Working 'intuition': What's my place in the world? The answer lies somewhere in the feelings that come up within me here in Upper Alton. These feelings point to something. They are significant, somehow...

Similar to Descartes' objective, in a very broad brush sort of way. Descartes was seeking intellectual foundations; Michael, personal foundations. For Descartes, the working intuition was that after a "general demolition" of his opinions he could find a point of certainty, something he knew without doubt to be true. His method was quite creative, really; he would subject his knowledge to the most withering attack imaginable. He would rule out whole categories of knowledge if they could be shown to include anything doubtful. The game was to see if anything survived. Something did: He could not doubt that he was doubting. Doubting was a form of thinking and at a bare minimum testified to the thinker's existence. Whence the famous formulation, "I think, therefore I am."

> After having reflected well and carefully examined all things, we must come to the definite conclusion that this proposition: I am, I exist, is necessarily true each time I pronounce it, or that I mentally conceive it.[7]

To be sure, Descartes wasn't the first to so conclude. Aristotle, in his *Nicomochean Ethics*, said that whenever we are conscious of our own experience of thinking, seeing or believing, we must be conscious of our own existence. But Descartes showed up at the right place and the right time: Europe at the beginning of the European scientific revolution. His work helped place questions of knowledge – what can we know? – at the center of modern philosophy.*

Descartes took this a step further. If "I exist, I am" was the undoubtable anchorage, what about the existence of the body? Of the whole world? Could it be doubted? The answer: yes! This is actually the central premise of a recent movie, *The Matrix*, in which the brain of (almost) every human is connected to a huge computer which feeds it false perceptions of the world. But since the world is usually taken as the

* By the way, the canonical phrase "I think therefore I am" does not actually appear in *The Meditations*.

example *par excellence* of solid, no-nonsense reality, Descartes's idea that the reality of the self was primary didn't fare too well. Nevertheless, he staunchly maintained both the reality of the self and its independence from the physical world; the self was the logically prior perceiver of the physical world. This claim drew strong critical fire from philosophers in the succeeding three centuries. How could a non-physical 'self' have any effect on a physical body? It was like saying there was a ghost inside a very complicated machine that 'really' made it do what it did, though the simpler operations of the machine seemed to follow normal physical principles without any ghostly aid. Was the ghost nothing more than the yet-to-be-understood complexities of the machine? Most modern philosophers would be inclined to think so and, consequently, most modern philosophy essentially did away with the ghost, declaring that we are really 'only' machines. *Very complicated* machines, machines that are far more complex than anything we can currently understand, much less build; perhaps even more complex that we can *ever* understand (though this is a minority position). Machines nonetheless – if one interprets 'machines' broadly enough.

> How very interesting that Descartes should fix on the knowledge of the existence of the self as the most certain of all knowledge! I'm sure he'd be distressed with its diminished reputation some three and a half centuries later: the self, reduced to the ghost in the machine. ☺

Michael laid *The Meditations* down and picked up *Pensées*. He'd once co-led a Pascal seminar at St. John's in which the limitation of reason had been hotly debated. But to find the exact quotes… There were hundreds of entries, from a few lines to a page or two. Fortunately, this Penguin edition had some rudimentary classification: Section One, XII: Submission and use of reason. Pages 83-85. That sounded hopeful. Bingo!

> Two excesses: to exclude reason, to admit nothing but reason.
>
> Reason's last step is the recognition that there are an infinite number of things which are beyond it. It is merely feeble if it does not go as far as to realize this.[8]

Blaise Pascal had already pegged reason as limited. Reassuring! But, oddly, Michael's first impulse was to wonder if Pascal might have sold reason a little short. If Pascal indeed took reason to be the sort of reason embodied in science – which he might well have, being a gifted scientist himself – then there was no problem. Michael and Pascal agreed. But couldn't reason be pushed a bit? Did reason really have to give up the game at the gap? The key step in the plausibility argument of the Flatland prisoner's argument to his jailer was that the idea of a mysterious 'third dimension' – whatever the hell that was – *rang true!* And it rang true to *reason itself!*

Reason can see beyond itself. Reason can see beyond reason. Paradoxical as that sounds, it seems to be true. The existence of the gap is based on solid science

– reason on steroids. Reality beyond the gap is by definition beyond ordinary reason, yet that does not undermine the scientific foundation of the major claim – the claim that there is a significant gap between reason and reality.

(...should I say reason <u>beyond</u> the gap, or reason <u>in</u> the gap? TBD. For now, let's go with beyond. ☺)

On impulse, Michael flipped back to the first page of his notebook:

<u>Immediate Problem</u>: Reduction of those feelings to psychological states (grief, problems w/father, etc.) or to character flaws (this whole enterprise is just self-indulgence, etc.).

He was grinning: that reduction clearly tended to rule out of court the possibility that the self was anything above and beyond its embodiment. The self 'itself' was an illusion, the ghost in the machine – perhaps the machine's software. All that was really required was a user's manual. And if the self was reduced to an illusion, there was no hope for anything as vague as a vision.

Of course! I've been tacitly assuming that any vision of my place in the world had to be totally reasonable – in the ordinary sense of 'reasonable.' I've fallen into the second of Pascal's excesses, to admit nothing but reason. More precisely, to limit myself to a feeble reason – namely, ordinary reason as the <u>only</u> reason.

...which neatly accounts for the painfulness of the censure I projected onto that guy at Horace Mann! Enfeebled reason would see 'vision quests' as pitifully quixotic. Childish attempts to deal with reality. To accept that analysis was to accept that my real task for this weekend was simple: Grow up. Ouch!!

Michael stopped writing, leaned back. The task for this weekend had been, in some measure, successfully completed. The unwarranted influence of ordinary anxieties had been identified and reigned-in; the immediate problem had been solved. Though he was still adrift, things were beginning to make sense; he was beginning to get his bearings.

But those ordinary anxieties had been reigned-in because they were ordinary; they were not implicated in the metaphysical difficulties of an epistemological gap. What about the gap itself? Here, surely, was a *wealth* of metaphysical difficulties!

Michael's initial euphoria was fading; he was beginning to harbor doubts. In fact, if one stripped away fancy words like 'metaphysical' and 'epistemological', was 'the gap' nothing more than a fancy way of saying there's just lots of stuff we don't know? How banal!

But upon reflection, that banality quickly showed itself a question-begging fraud. The phrase "just lots of stuff we don't know" implied (by innuendo) that the only stuff to be known was ordinary and thus – if ordinary – available in principle to ordinary reason. Which, of course, just begged the question: *is there* 'stuff' not available to ordinary reason?

But the primary implication of the existence of the gap was that there was almost certainly inconceivable stuff. *Literally* inconceivable. It is almost impossible to overstate the sheer magnitude of the disparity between reality as it is and the reality we could possibly grasp – even given tens of thousands of years of knowledge accumulation. (Still a geologically puny time scale.) Think of the gap in a human's comprehension of geometry and the comprehension available to a cat – even a cat as smart as Sam. Now, replace a cat with a spider. Now replace the spider with a Cambrian trilobite, wandering the floor of a shallow ocean, half a billion years ago. The disparity in comprehension defies description!

To object that the brain of a trilobite was not capable of anything comparable to human comprehension would be to miss the point: The 'conceptions' of a trilobite stand to human conceptions – for example, the human conception of the world as inhabited by animals with brains – as human conceptions stand to something far beyond us. As Eisley said of the mammal whose skull he chanced upon in the slit, "[It] had never lived to see humans, and I, what was it I was never going to live to see?"

Yet, that doubt was no sooner stilled than another arose: Even if things beyond the gap were incomprehensible, might they not therefore also be inconsequential? Why did these things have to play any role in our life at all?

> Everything on this side of the gap would be, by definition, available (eventually) to reason. If we were somehow shielded from anything beyond the gap, then there would be no incomprehensible, inexplicable realities for us. We'd effectively be Flatlanders blissfully unaware of space...

Michael leaned back and stared out the window. Watched several cars go by. This shielding was just ... suspicious!

> This just doesn't sound right! Why should we expect to be insulated from the inexplicable realities implied by the deep space, deep time argument? Isn't this just 'what you don't know can't hurt you' writ large?

> ... but if we aren't insulated from those inexplicable realities, where are they?

Michael looked around the room. There was nothing that felt inexplicable. Everything seemed perfectly ordinary. No sign of any indistinct, ambiguous 'something.' If anything, it was so ordinary it felt just a bit boring ...

... but perhaps words like 'ordinary' and 'boring' were offering subtle hints. The real problem wouldn't be actually *seeing* something inexplicable, it would be seeing something *as* inexplicable – not just ordinary. If things-inexplicable presented no threats, we would soon cease to see them – or at any rate, to pay them much attention. We would take them for granted. They would quickly become just *ordinary*.

Looking again at the room, and given today's ruminations on the self and the world, the subjective and the objective, a thing-inexplicable immediate appeared. Lots of them, in fact. People! The room was filled with selves-with-stories. In fact, these interior selves were arguably the real referent for the word 'people', rather than the objective bodies that one actually saw. Nothing in front of him had changed, but the room now felt ... a bit spookier. It was full of ghosts-in-machines!

> All these selves... Could it be that Descartes unwittingly stumbled into the gap? The problem of resolving the tension between the self and the world was one of the chief problems that subsequent philosophers would beat him over the head with. Could the real problem arise from their dogged attempt to explain the inexplicable? To use normal reason outside its normal operating limits? To apply it to a case of an inexplicable reality? Namely the irresolvable tension between the objective and the subjective! **

** Thomas Nagel, a well-known New York University philosopher, wrote an entire book dedicated to the tension between the objective and the subjective: *The View from Nowhere.* "This book is about a single problem: how to combine the perspective of a particular person inside the world with an objective view of that same world, the person and his viewpoint included. [In other words, a unified worldview. But] instead of a unified world view, we get the interplay of these uneasily related types of conception, and the essentially incompletable effort to reconcile them."

Nagel, Thomas. 1986. *The View from Nowhere.* New York: Oxford University Press. p 3-4.

Michael leaned back, put his pencil down and stretched. Then got up and stretched more. It had been a long day. He looked at his watch: 3:30. He was due at Roy Greywood's for supper at 4:30 – early even for a Midwestern supper, but there was still plenty of time to get home and change. He picked up Descartes and Pascal, walked back to the stacks, then stopped. A sign on the shelf politely requested, PLEASE DO NOT RESHELVE BOOKS. How nice. But Michael figured he's shelved these things thirty years ago, and he could damn well shelve them today. So he shelved them, left the library, and walked to his car. Grinning.

Driving down the last three blocks of State Street, he turned east on Broadway and began the long climb back to Upper Alton. In a few moments, he passed the Alton Belle, docked on the Mississippi, to his right. He glanced pensively at the boat, not entirely comfortable with a gambling casino as a conspicuous part of Downtown, but… Well, Alton wasn't his town anymore

Michael passed the Wedge Cafe and Springmans, coming finally to the river-level terminus of Washington Ave, and began the steep climb to Upper Alton.

THE EVENING OF THE SECOND DAY

Saturday, April 4, 1992

Roy Greywood had known Michael and his mother for many years. In 1963, after ten years in her parent's house, Virginia complained that she was caught between the generations – her aging parents on one side and her teenaged son on the other – so she found a small duplex apartment on Main Street where she and Michael lived until he left for Purdue in 1966. Several years later, she moved to a small but comfortably furnished, two-bedroom apartment that Roy Greywood had set up on his second floor; the Greywood children had left home and Roy liked the idea of putting the top floor to good use. Also, he and his wife had known Virginia for many years; the rent they charged was so low they could probably have claimed a hefty tax deduction for charity. However, after living ten years upstairs from the Greywood's, Virginia faced a new problem; her parents were no longer able to take care of themselves. After agonizing whether she should move in and become a more-or-less full time caretaker, she decided the task was really beyond her. Reluctantly, she was forced to find her parents a nursing home. Michael returned to Alton to help move his grandparents to the Eunice Smith nursing home, then helped his mother move out of the Greywood's apartment and back into the family house on Washington Avenue.

As a boy, Michael had been a little wary of Mr. Greywood. Roy's natural expression was austere and foreboding; he seemed always on the verge of delivering a stern Presbyterian lecture. However, returning to Alton years later as an adult, Michael saw beneath the surface expression to a man who was vigorously committed to the welfare of the people around him. Active in several local charities, Roy made a special effort to help his peers. The widow next door to him, for example, was several years older than Roy and very frail. To allow her to stay at home, he did her shopping, took care of errands, and had even hooked up an alarm she could trigger that would sound in his house, in case of emergency. Michael liked to think of Roy as the sort of man typical of Alton. Romanticized, probably, but an agreeable image nonetheless.

Michael arrived at Roy's house a few minutes past four-thirty, hoping the eighty-year old Roy would let him drive – and a little guilty at being even a few minutes late. The house looked pretty much like any other in Alton. White wood paneling, a full length front porch, a porch swing, and a small but well-manicured front yard; a modest middle class home, directly across the street from Alton High School. Roy was waiting for him at the front door. "Well, Michael! Welcome! Good to see you! Come on in!" Michael stepped inside.
The front room was tranquil, with an atmosphere of time suspended. The furniture was rich, dark mahogany; the walls were deep yellow, covered with carefully arranged pictures. A large photograph of Roy's wife looked down from the far wall, above a well-kept fireplace; she had died in the early eighties.

Roy motioned Michael to a couch. As he sat, his eyes adjusted to the dark and the pictures that surrounded him came to life. As he asked questions, Roy expanded on the photographs of his family, telling a little tale about each one. Working his way around the room, a whole world opened up: graduations, marriages, children, grandchildren, adopted children. Roy's features were quite animated – much more animated than Michael remembered. Depths opened up, connections to the people and the photographs that left Michael with an odd sense of closeness to Roy. Somehow, Roy's relation to all these people from the past made him more real, more present here and now.

Roy stopped for a moment, opened an end-table drawer, and sorted through several sets of snapshots. "Ah ha! Here it is!" He sat down next to Michael and handed him the photograph. It was a picture of Roy with Virginia and two other people, taken downtown near the river. "The summer before your mother died, she and I, along with some friends of mine, took a real nice trip, up the river to Hannibal. Real nice." Roy laughed, "I had to twist her arm to get her to come, but I finally got her to agree. When we all got back, we got a passerby to take our photo and promised we'd all get together again." He paused for a moment. "Didn't work out." He paused again, this time just a shade longer, as if considering how much to say. "Michael, your mother was a good woman. Virginia had it pretty hard. Single mothers were unusual when she came back to Alton, not like now. She had to make it by herself, with her parents to help her.

Your mother talked to me a couple times. I don't think she talked to many people about it. You know that things got pretty bad there toward the end of her marriage..."

Roy was abruptly silent; he glanced at Michael and then down at the floor. "I'm sorry, Michael, this is none of my business!" That stern Presbyterian look had returned, but now it was directed against Roy himself.

"Don't be sorry, Roy, please. I'm glad she talked to you. She really didn't tell me much about it, and I remember next to nothing about those years, but reconstructing it as an adult, I've pieced together that things were pretty bad at home. Not physically so much, but psychologically. I would really appreciate hearing what she told you. That is, whatever you think she'd want me to hear."

Roy glanced down at the photograph. "Thank you, Michael. She was real proud of you, you know." He paused briefly, and then continued, "I'm not surprised she didn't talk to you much about your father. He was a real smart man, all right, but unstable. Liable to do anything, without warning. When she got to talking to me, I felt like it was the first time she had ever really opened up to anybody about all the things she went through."

"She told me about the day she arrived back here in Alton, just down the road here, at College Avenue Station. I think maybe it was the C&A back then, the Chicago and Alton

Line, before they got taken over by GM&O. No, wait . . . I think that was just after the war…" Roy shook his head and returned to his story, "Well, anyway, her father was there waiting for her. A friend from church too, as I remember. They gathered everything up and went back to the house. I think maybe it was later that day, she was sitting upstairs in her bedroom – the same room she'd had as a girl – and it all just started to hit her. She'd gone out into the world, worked, gotten married, had two children, seen one die of crib death – I think they call it Sudden Infant Death Syndrome now." Michael nodded, "His name was David. He died before I was born." Roy continued, "Well, sir, here she was, back home, back in the same room, but now with a small boy to raise. That was a real black day for her. But Virginia was determined she would raise you right. She got that job at Springman's and just set her mind to it. Fortunately, she got along OK with her parents, and they were real supportive. Later, when you left home, and her parents got to the point where they couldn't take care of themselves, she was there to help. Not easy. I can tell you that. Not easy. No, sir. Not easy at all."

That seemed to trigger a memory. "Just a minute, Michael." He leaned back to the end-table and sorted through the photos again. "Here it is," he said, handing another photo to Michael, this time a much older, black and white one. "I gave your mother an old photo of my family once. One I'd made several copies of. Well, of course, she had to go right out and make a copy of one of her own photos to give me in return."

Michael recognized the picture. He had the original now back in his attic in Maryland. It was a wonderful photo, taken sometime in 1918, of his mother, grandmother, great-grandmother and great-great grandmother. His grandmother and great-grandmother were standing behind a sour-looking old woman – his great-great grandmother – who was holding the infant Virginia. "God, Roy, I remember this! That's Grandma and her mother, Eva, behind my great-*great*-grandmother!" Michael paused for a moment, thinking, "I can't recall her name, but I seem to recall she was born in 1838. I guess this photo must have been taken on Eva's family farm, back in Ohio."

Roy continued, "When your mother gave me that photo, she told me a little story about it. She'd always liked that picture, so when her mother was at the Eunice Smith nursing home, just up the road, she framed the original and set it on the table next to her mother's bed."

Wednesday, March 3, 1982

Virginia's mother had died in her sleep. The family had come home. Attended the funeral. And left. Now, Virginia was boxing up the few things in her mother's room at Eunice Smith. She put the card from the staff into the box. Everyone, from the director on down to the cleaning staff had signed it; the small card was covered with signatures.

The last thing Virginia picked up was the photo next to her mother's bed. It was a picture of her mother, grandmother, and great grandmother, back at grandma's farm in Ohio. Virginia was on her great grandmother's lap, less than six months old. She didn't remember anything about her great grandmother, who had died somewhere in the early twenties. She did remember, though, what her younger brother had said about the sour expression their great grandmother seemed to have in all her photos: "Old women eat crab apples! That's all they eat! Every day, they gotta have their crab apples!" That had always exasperated Virginia – which was, of course, exactly why her brother persisted in saying it.

Now, looking at the photo, she thought to herself that, in fact, old women don't eat much of anything. She had spent hours trying to get her mother to eat, with little success. No longer.

She put the photo in the box, closed the top, and left.

⚜

Roy and Michael looked at the photo for a while. Then Roy stood up, looked down at Michael and smiled, "Well, we'd better get going. I got reservations for us at the Midtown Restaurant." Michael rose as well, handing the photograph back to Roy. "OK. Sounds good. I haven't eaten there in a long, long time. I have my car outside, so I can drive if you like . . . " A look of surprised indignation came over Roy's face, and Michael instantly backtracked, ". . . unless you'd prefer to."

Roy's smile returned, "No, I'll drive, Michael." He moved to put the photograph back in the end-table drawer, motioning for Michael to follow him. "Let's go out the back way. I'd like to show you my new squirrel feeder." He followed Roy out the back door, down three steps, and into the back yard. Several bird feeders were attached to the house; one hung from a large oak tree. As the two of them neared the garage, Roy pointed out his newest creation: Two dried corn cobs on wooden dowels, carefully placed on a platform secured to the side of the garage. "Helps keep the squirrels away from my bird-feeders. At least, that's the idea." Sure enough, the contraption was an easy squirrel-jump from the oak tree, but far enough from the ground not to become an easy raccoon feeder. And as far as Michael knew, cats didn't give a damn about dried corn, so it wouldn't become a cat-feeder either.

Once inside the garage, Roy opened the car door for Michael. The car was larger than anything he drove, back in Maryland. He had grown accustomed to compacts and subcompacts; this felt a bit strange. This was how cars felt thirty years ago. Big.

The Midtown Restaurant felt warm and familiar; in a far corner, a man in a dark suit played pop tunes on an organ. The one thing that felt new was Roy's suggestion that they split a half carafe of wine. As they ate and drank, Michael prodded Roy to talk about his own boyhood in Alton, which he did enthusiastically. These were all old stories, but they were new to Michael, who listened with fascination to stories of an Alton he had never known: a train station downtown when trains were the standard way to get to St. Louis and street cars were the way to get around town; the building of the Lewis and Clark bridges in the late twenties; the kidnapping of a prominent businessman in the early thirties, whose safe return was announced by the only 'extra' edition of the *Alton Evening Telegraph* ever published. Roy even cleared up a puzzle for Michael: the real story behind a boat wreck. In the fifties, there had been a small boat that had sunk just downstream of the Clark Bridge. As a boy, Michael had seen the sunken boat and, of course, envisioned skeletons and treasure. Roy told Michael the boat had belonged to Chicago gangster Al Capone! An Alton man had purchased it when the estate of the gangster was being settled. The boat was brought down the Illinois River to Alton, but something evidently happened and it sank below the bridge. Scavengers had removed all the salvageable parts. Time and the river had done the rest. By the time Michael was in high school, the boat was gone.

Roy was a lifelong and active member of the Masons. Starting in Junior High School, Michael had been a member of DeMolay – sort of a junior version of the Masons – but abandoned the association once he left for college. As an adult, neither he nor anyone he knew was a Mason; he'd simply moved into a different culture that didn't include the Masons or anything like them. One of the staples of Masonic life was elaborate ritual that had to be memorized, word-for-word. As a DeMolay, Michael had struggled to memorize comparatively short speeches – with mixed success. Roy, however, had risen about as high in the Masonic Order as a man could go, so he'd become highly adept at mastering the spoken word. With that mastery, he had discovered the world of poetry and enthusiastically absorbed countless poems. He reeled off several, then paused.

One of his favorites was entitled, "The Pleasant Mile," which he recited for Michael:

People come into our lives
And walk with us a mile.
And then because of circumstances
They only stay a while.
They serve a need within the days
That move so quickly by,
And then are gone beyond our reach.
We often wonder why.
God only knows the reason that
We meet and share a smile –
Why people come into our lives
And walk with us a mile.

When he finished, he paused again then said, "Michael, your mother, Virginia, came into my life for more than a mile, and I'm eternally grateful."

After dinner, Roy parked in his driveway rather than his garage. He would be going to a meeting at the Masonic Temple; *that* had been the reason for the early dinner. As they got out of the car, Roy said, "Oh, yes. There's some mail for you. Let me get it." He disappeared in the house, and returned with a few small envelopes. "Brian, down at the Wedge Bank, knows I've been keeping an eye on your place. I was down there last week, and he pulled out a small packet of mail. He would've just forwarded it to you, but I said as how I'd be seeing you. Told him I'd just give it to you face-to-face. Never know about forwarded mail. Anyway, here it is."

Michael smiled as he pocketed his mail, wondering just how many people in Alton knew Roy. He probably knew most of the bank staff. Made sense they'd use him as a sort of an informal courier. "Thanks for the mail. I enjoyed dinner. I've never heard some of those stories. Like about that boat at the foot of the bridge. So it was Al Capone's! Well, again, thanks." Roy shook his hand, "Thank *you* for coming by, Michael. Keep in touch now!" Michael replied, "I will," and thought to himself, with a feeling of guilt, 'I *hope* I will.'

Michael wasn't very good at keeping in touch with people. Driving off, he thought about that. It wasn't that he always failed to keep in touch – he had set up this dinner with Roy, after all. It was just . . . It took energy to reach out. He thought about the people who had reached out to *him*. First, of course, there was family: Mom and Grandpa and Grandma. Years later, after moving to Baltimore, Aunt Mae and Uncle Merle had reached out in countless ways. Beyond family, there were many people in Alton. Teachers, friends. On and On.

Nearing home, he passed the site of Bernie's old house. "I'll call Bernie," he said aloud, then chuckled quietly to himself. Impulsive contact was still contact, right? He still got points for reaching out, right? Ten minutes later, he was on the phone.

No answer. The phone answering machine kicked in, "Hi, this is the residence of Theresa and Bernie Kohler. After the…" Michael hung up. He didn't want to leave a message. What would he say? Maybe he'd try again tomorrow.

He leaned back, let out a long breath, and tried to remember his last conversation with Bernie, at the funeral home.

Sunday, November 4, 1990

THE REVEREND KENNETH L. CHANDLER, a retired minister with the Assemblies of God, died on Thursday evening, November 1, 1990, at Alton Memorial Hospital. Mrs. Virginia M. Howard, elsewhere in the same hospital, died early the following morning. Both were scheduled for visitations at Elias Smith on Sunday afternoon. A sign at the front door of the funeral home directed the Reverend Chandler's numerous flock to the larger main viewing room; those attending Mrs. Howard were directed to a smaller room off to the left. Before her room was opened for the general viewing, Michael was allowed to spend a few minutes alone with her. He stood quietly by his mother's casket, once holding her hand. Then, very deliberately, he walked back out to the main foyer and assumed the duties of a surviving son.

Michael was pleasantly surprised at the number of people who arrived. Since his mother had been somewhat reclusive, he figured few people outside the family would come; he was wrong. Neighbors came. People from Main Street Methodist came. Her boss Jack Springman and several colleagues from work came; Jack made a point of telling Michael how utterly trustworthy Virginia had been. All of the women from her sorority came; at one point, they gathered around the casket and recited a ritual. Michael wondered to himself if fraternal associations with memorized rituals were typically Midwestern. The ceremony felt very different from anything in his current life on the East Coast; it also felt familiar – and very comforting.

Several of Michael's old acquaintances from Alton had evidently seen the obituary in The Telegraph, and came as well. Lee came, a good friend from his DeMolay days. As they talked, Lee made a wave-like gesture with his right hand; that gesture instantly transported Michael back to their high school days together, when they would walk home together from the Masonic Temple on Washington Avenue, and Lee would use that same gesture to make a point. The biggest surprise, though, was Bernie Kohler, whom he hadn't seen for over twenty years. He hadn't even known that Bernie still lived in the area. They sat down together and tried to catch up quickly on two decades.

Things had become briskly interesting for Bernie shortly after high school; falling three credit hours behind his class in college, he came to the immediate attention of Uncle Sam, who sent him a 'Greetings and Salutations' letter inviting him to join the Army. In an initial battery of Army tests,

Bernie had scored especially well in electronics; with the help of a friendly recruiting sergeant, he picked the electronics specialty with the longest training school on the theory that nobody shot at you there. That specialty turned out to be radar repair. What the recruiting sergeant failed to tell Bernie was that the radar he would be trained to repair was used for tracking incoming enemy mortar shells – a radar in wide use in Vietnam. Out of a graduating class of thirty-five, thirty went directly to active combat units. Bernie was one of the five who were detailed to a civilian depot in Sacramento, where he became a military pall bearer for his unlucky peers. A year later he was reassigned. Again, luck was with him; he went to Germany rather than Vietnam. Years later, he married and now had two boys. As Bernie left, they agreed to stay in contact.

Michael talked to many people. Some were friends of his mother whom he had known; some were friends he hadn't known. One woman in her early twenties introduced herself as his mother's hair dresser! She had taken care of Virginia for over a year. Of all the people Michael met, he was struck the most by this woman. Not because of anything in particular that she said, but for the sense it gave him of the complexity and depth of his mother's relationships. Listening to this woman, it was evident that Virginia had been much more outgoing than Michael would have supposed. He felt very good about that.

Later that night, back at the house, surrounded by friends and relatives, Michael withdrew a little from all the talk, closed his eyes and just listened to the conversation swirling around him. In one instant, many different tales were spinning out

"You *bet'cha* he did! Well, sir, the next day. . ."

". . .and what that poor girl had to put up with!"

"But, don'tcha *see? That's* what they were really *up* to. . ."

". . .Well, *I'd* never let any man get away with that!"

"Oh, sweetie, while you're up could you get me a Coke?"

The enormity of the associations and connections of just the lives in this room! Normally, Michael would have shied away from such a buzz of conversations. That night, though, the whirl of conversation was deeply consoling.

THE NIGHT OF THE SECOND DAY

Saturday, April 4, 1992

He opened his eyes and looked toward the living room, where all that conversation had taken place. Then leaned back in the chair by the phone, feeling suddenly tired. It had been a very full day. The setting sun was streaming in the kitchen. A bird chirped away steadily somewhere in the back yard. A robin? Maybe. Its sing-song chant echoed among the houses, punctuating the enveloping silence. The light began to withdraw slowly to the west. The colors in the dining room faded, became less distinct. A picture on the wall transformed itself into a small black rectangle with just a hint of something within, as if it were now a small window into some other universe, some other dimension. Something there, but impossible to see. The yellow kitchen walls dimmed to a pale gray; the green dining room walls, to a darker gray. The dark brown dining room furniture turned darker. Finally, the robin was silent.

Michael watched the dining room table, but without fixed attention. As the light faded, the house began to recede, to merge with the sound of his breathing. After a while, all that seemed to remain was the sound of his breathing.

Abruptly, the timer-driven lamp in the living room switched itself on. Startled, Michael looked at his watch. 7 p.m. He looked back at the dining room table, which had sprung up suddenly from the deepening dusk and taken its familiar place among the furniture of a domestic evening. The walls of the dining room were again green. The kitchen, yellow. And with the definite colors, so also the definite shadows.

He stood and walked to the kitchen. Though still full from dinner, Michael figured a little dish of ice cream would be good. Opening the unopened pint, he spooned a little bit. Not great, but not bad. Sitting at his accustomed chair, with his back to the back wall, and the door to the basement at his right, he worked out a few rock-hard spoonfuls. Maybe not so good after all. Finished, Michael looked vacantly ahead, through the house and out the front window. No traffic outside.

A slight disquiet nudged him to look down the stairs to the basement. Surprised, he now felt definite anxiety. Fear, in fact. It had come upon him rapidly, but also with the feeling it'd been there all along. That he'd simply failed to notice it. He was actually afraid to go downstairs. Earlier that morning, as he came downstairs from the second floor, he'd felt a twinge of regret that the night before – when he had that uneasy feeling at the top of the stairs – that he'd not gone down. To what? To face it? What was 'it'? What would 'facing it' consist of?

Still uneasy, he determined not to make the same mistake twice. Flipping on the light, he descended to the basement. He walked by the table, ducked around the furnace, and completed a circle back to the foot of the stairs. Then looked underneath, to the place where spaceships and haunted houses had once lived. He half expected something to be there, but whatever it was kept out of sight, vanishing when directly sought. At the top of the stairs, it had felt like he would feel more fear when he came down. He didn't.

Not sure what to do, he recalled the mail Roy had given him, now stuffed in his back pocket. An old lounge chair sat just at the foot of the stairs; the light hanging from the ceiling would make for easy reading. The chair was covered by one of his Grandmother's quilts – a quilt that was ripped and stained, and so had not been taken by anyone. Back in Maryland, he had probably half a dozen quilts. Another, especially so shop-worn, wasn't worth it. He retrieved the letters from his pocket. It had begun to rain outside.

Scanning the envelopes, one immediately caught his attention, addressed to "`Mike Howard/Please Forward.`" The return address was "`Gladys Poland, 53rd and Woodward, Elgin, Illinois.`" The letter had had been typed on an old, mechanical typewriter. The individual letters were uneven; the typeface was evidently worn. The word-choice was vexing: Michael never used "Mike." He didn't like the name.

A premonition crossed his mind as he tore open the envelope, fulfilled when he looked at the letter and instantly spotted the phrase "commit suicide."

Saturday, February 29, 1992

SLEEP, GONE. EYES ABRUPTLY OPEN. The room still dark. Black night. The grandfather clock, slowly chiming. Guy counted the chimes:

One. Two. Three. Four. He desperately hoped for the chimes to continue. More chimes meant less night left to endure. Four wasn't too bad. Only a few hours till dawn. Nights were horrible, unbearable. He couldn't stand another night.

Mabel had been in the hospital nearly a week now, after suffering a serious stroke. Guy knew there was no hope for him without her. He had tried to suppress that knowledge, but it periodically erupted in waves of panic. He lay awake at night, sweating, shaking. With nothing to stop them now, the voices were swarming in like vengeful demons. Guy could hold his own against them with bursts of fury and hatred, but it was a hopeless defense. He couldn't hate long enough or hard enough. Exhausted, he would drop into dreamless sleep. Then, suddenly, he would snap awake.

Five. **Six**. Headlong hope. Six o'clock was nirvana! A whole night's sleep! It was wonderful! Extravagant!! A whole night, not broken into desperate shards. Rest. Respite from the gathering horror.

He was alone in the apartment. Without Mabel, his daily routines had been fatally disrupted. Those routines, like the binding of an old book, held his days together. Without Mabel, the binding lost its remaining strength. The days fell apart. Disjointed. Disordered. He would watch TV, turn to ask Mabel for a glass of milk and abruptly stop. Mabel was not there. Mabel was gone. Guy would grasp angrily at the remote control and turn off the TV in a burst of spite. Since Mabel had been taken away, he had not been able to watch a single complete program. The voices would tell him there was nothing left, that this was the end, that his life was finished. Finally, fatally broken.

He would long for sleep, for some kind of release. Finally, he would fall into the blank sleep of exhaustion. But once the bare physical necessities were satisfied, once his biochemical constituents revitalized themselves, the dreams would return. Demonic. Threatening. He would snap awake. He had descended into a perpetual twilight. Awake, he wandered from one thing to another, each increasingly disjointed. The voices would swarm in. He would fight them off. Physical exhaustion would slowly, mechanically increase till, in a moment of calm, he would drift into sleep. But then his body, responding to the dream-threats in this permanent dusk, would shoot cold adrenaline into his arms. He would jerk awake. Shaking.

Seven. Seven? Seven o'clock? But it was still dark…

Eight. …Eight o'clock? Was it cloudy? Was it raining?? He froze.

It was **eight** o'clock **at night**! Adrenaline flooded both arms. Cold. Fear. The shaking abruptly returned. Worse than ever. Worse that it had ever been. His whole body shook convulsively. It was still night! Early night!!

Nine. **Ten**. **Eleven**… Twelve? Twelve? Where was twelve?? Desperately, he clung to twelve, willed that next chime, the chime that would at least bring him into the next day… He brutally forced himself to concentrate on that. If it were just tomorrow… A new day… He rolled out of bed, grabbing for the walker. Without treatment, his arthritis had degenerated terribly. Painfully he willed himself up, leaning heavily on the walker. The pain and the effort helped focus his attention. The panic subsided. He headed for the living room, to check the clock. He must have missed a chime, missed a count. The count was twelve, not eleven. It had to be…

Beyond the end of the bed, the back left leg of the walker snagged on the rug. As Guy shoved it forward for the next step, the walker twisted violently to the left, sending him crashing to the floor. His joints shrieked pain. Guy howled his fury and frustration, striking the treacherous walker. Shaking with rage, he crawled on through the short hallway, then into the living room. He raised himself on his right arm, the elbow and shoulder joints wailing pain. Brutally, he wrenched himself up, deliberately, spitefully sending a further wave of pain through his traitorous joints.

The light switch was a dark, vaguely rectangular shape against the dully glowing white wall. He grunted, trying to stretch his left hand up to the switch, but the added burden to his right shoulder was too much. He crashed into the bookcase. Instantly, there was a slipping, shuffling sound. The pile of books on the top of the bookcase poured down on him. A sharp, hard pain hammered his skull. He breathed in sobs. He reached up with his left hand to the top of his head. Something wet. Sweat? Panting, he stared into the room. Dark, unintelligible shapes loomed over him. He rolled slightly and, with his left arm, shoved himself to a sitting position, his back now to the wall, the switch above him. Extending his left arm up and behind, he probed blindly. There it was. He could feel the light switch. Seizing the brief flush of strength from this little victory, he jerked his body up. Again. And again.

The light came on. He squinted. His arm fell back down. The shapes in the room, illumined by a single light, studied him. The table and the chairs stared down scornfully. Guy lowered his eyes. He looked at his left hand. Blood. Recognition dawning, he touched the top of his head again, and felt a small but sharp answering pain. He brought his hand back around. It was covered with more blood. Now conscious of his injury, he could feel the dull, throbbing pain; a new injury, it had not yet begun to really hurt. Among the books on the floor in front of him lay a large, bronze gargoyle bookend. That was what had hit him. Breathing heavily, Guy looked at the gargoyle. It looked back. Examining him.

Guy noticed the sound of his own breathing. As his attention turned toward the sound, he became aware of the quiet ticking of the grandfather clock, to his right, just beyond the door. He turned his head slowly, looking at the clock's base. The pendulum moved back and forth. Slowly. Deliberately. His reflection in the spherical bob of the pendulum was like the black pupil of a single eye, shifting back and forth, its inhuman stare fixed on him. Back and forth. A lidless, unblinking eye. Staring. Waiting.

Guy looked up to the face of the clock. He saw the hands, pointing in a familiar pose. Two unequal hands cradling the number twelve. It was some time after eleven. Not twelve. Eleven. Still night. Saturday night. An hour short of Sunday. He jerked his left arm up viciously, back against the wall to the light switch. The room plunged again into darkness.

He began to sob, venting an emotion so dark it had no name. Blackness, a despair so dismal there was an exhilaration to it. A dread so profound it seemed to reach up to him, beckon him, pull him further down. Down into a pit of terrible darkness. Guy cried out as he swung himself around, back through the door. He rose on his hands and knees, crossed the hallway, crawled into the bathroom, jerking like a puppet yanked by some demonic puppet master.
Numbed, filled with a fiendish energy, he lifted himself onto the toilet, then up to the washstand, and then reached to open the medicine cabinet, feeling for a razor. Medicine bottles rained down on him, some shattering in the sink, some on the floor.

He felt the handle of the razor and, with a vicious shove, pushed himself backwards, briefly onto his feet. Standing for the last time. But with no fixed intent to stand, he twisted left and lunged for the bathtub, falling into the shower curtain. Ringing, staccato shots echoed off the walls as the curtain tore loose, breaking his fall into the tub. He fumbled under the curtain, locating the plug, and shoved it firmly into the drain. The walls echoed again, now with the sound of pouring water, splashing loudly into the steel tub. With a final, mortal effort, Guy settled in. A sharp pain made him cry out; the gash on his head had struck the metal faucet. He lay in the tub, panting, sobbing as the water gushed in around him. He clutched the razor in his right hand. The tub filled. He closed his eyes…

After a few minutes, water began to enter his nose. Coughing, he jerked up, reaching behind him to turn off the faucet. The sounds in the bathroom ebbed promptly, diminished to the swishing of waves reflecting back and forth across the tub, against his body, and against the shower curtain that had grotesquely wrapped itself around him. The water covered all but his head and his angular, bony knees.

A decision point had arrived. With a sudden burst of hate, he screamed and slashed at his wrist, again and again, overpowering the pain and the horror with pure, malevolent spite.

His energy spent, he stopped. Both hands splashed back

into the water. A dark plume of blood began to envelop his left hand. Suddenly, he began to cry. But not the dark, black, mindless, nameless tears of a few minutes ago. He began to cry the cleansing tears of pain and release. The frantic tears of an infant, yearning for comfort, for his place, for his parents. The compressed years of tortured, terribly lonely pain burst out. All trace of willfulness vanished. He cried, "Please! Help me! Please! Help me!"

His energy ebbed further. The cries subsided to a whisper: Please. Help me. Please…

In the soundless pause following this release of clean pain, a quiet voice urged him to open his eyes, to see the blood, to bring himself back from the brink. Even now, that gentle voice endured in this dark landscape. But still he resisted. He would not turn back. He would not open his eyes. He could not open his eyes…

The grandfather clock continued to tick. The pendulum continued to swing.

Finally, midnight. This time, the clock chimed twelve times. But Guy was no longer in the apartment; he had just been taken to the emergency room of a nearby hospital.

Yesica's parents, Marie and Hector, lived next door. They had taken Guy and Mabel under their wing, appreciative of the kindness shown their daughter – now away at Northern

Illinois University in nearby DeKalb. They had coaxed a key from Guy, 'just in case.' They were both in bed when Marie thought she heard a sound. Was it Guy yelling? She knew of Mabel's stroke, and thought about checking but … well, Guy was Guy and unpredictable. Marie and Hector considered the situation and decided it was worth the risk. They'd both had plenty of interactions with Guy and had learned to brush off his peculiarities. Hector went to Guy's door and knocked once. Twice. Then used the key.

March 4, 1992
Elgin, Illinois

Dear Mike Howard,

My name is Gladys Poland. I am a good friend of your stepmother. We have been members of the same church for many years now. You probably don't remember me, but I met you when you were here some years ago. You seemed like a nice man. Your father talked about you a lot. Your stepmother had a stroke. She is not doing so good.
Last weekend, your father tried to commit suicide. The Mexican couple next door heard something and found him before it was too late. Thanks be to God. Thought you ought to know so stopped by your father's apartment to get your address. Couldn't find your address, but found your Mother's address. Hope this gets to you. If I find your address later, I will send this again. I remember that you live somewhere on the East coast.

Sincerely,
Mrs. Gladys Poland

Michael had misread the note! His father had *not* committed suicide – or at least not succeeded. Michael felt relieved, but in a detached way. As the minutes passed, the feeling registered itself more as numbness. He looked around the basement for a while, fixing finally on the furnace. In the background was the subdued sound of the rain outside, like the static from an old radio. He continued standing for a while, then sat down.

He thought about the last meeting with his father. It seemed to end so well, with that exchange about the record – though even now, if Michael saw an old 78 record with a purple Capital Records label, it still struck a deep nerve. But that had been cleared up, hadn't it?

He remembered a moment, in their conversation, when his father had started to get visibly angry. Shaking. Red in the face. Michael had felt something just then, but forged ahead with his analysis on the story-like nature of people. Hadn't that feeling been released later? His father in the chair by the TV, and Michael standing beside him? Hadn't Michael's outburst about the record-breaking ventilated his anger? Hadn't his father apologized?

He reflected on that moment, he and his father sitting in the two chairs in front of the TV in Elgin, Illinois, on a hot, August evening, Mabel fixing dinner in the kitchen. The moment when his father had begun to get mad... No, not mad. Furious. Maniacal. Demonic. "Dad is mad" might be adequate for the child-Michael, but not for the adult-Michael.

Yet even this more sophisticated analysis was not truly adequate. What was going on exceeded all human analysis.

He thought of his father. Of the terrible, frightening pain that had been with Guy his entire life. Desperation so deep, so dreadful, so horrific that it was beyond the reach of sane imagining. A life lived on the edge of Hell.

Michael continued to reflect on that moment, the moment his father had been visibly angry. Then, startled, Michael put his hand to his stomach. Was he actually getting *sick*? Had it been the ice cream? Had it gone bad? He laughed. This was ridiculous. People only got sick like this in movies. But even as he thought these thoughts, he knew he really was getting sick. That moment when you realize the discomfort you have been feeling is tinged with nausea, and has been with you for a while. You had subconsciously hoped it wasn't, but it was.

Michael began to breathe more heavily. He felt sick, but in an extraordinarily detached way. As if he were experiencing someone else's pain. He noticed a paper bag near his foot, picked it up, put it over his mouth. But in such a bizarre, disembodied way! The image of the bag had triggered the thought, "Try to vomit, you'll feel better." He was trying to help himself. As if he and himself had detached, as if there were now two Michaels in one body, one helping the other.

He realized he was shaking. Incoherent thoughts and feelings poured though him. He forced that stream into words

– words marking the border at which the deep, inarticulate things inside him reached up for expression. But expression was only possible with words. Words that took him outside himself, to some place where this nausea was not the measure of everything. Not an unbounded agony, but bounded, connected, tied to something, to someone. Not the hopeless reaching up of one utterly alone, for whom reaching was fruitless. A reaching for what? Alone, who was there to reach for, there being nothing but a lone human self?

Dimly, Michael knew his hope lay in continuing to force out words. Fragments. Sometimes just sounds with the semblance of meaning. One word that began to recur was "forgive."

"God, I forgive you Dad! I really do! God, I really do!"

The reaching had transformed itself into a reaching-for-Dad. Not to get help, but to give help. To reach back down and pull Dad up out of Hell. To let Dad know that he was really, truly forgiven.

Not long after Michael's words reached the God-I-forgive-you-Dad level of coherence he realized he had stopped shaking. Not just shaking less, but not shaking at all. He was still breathing hard, but the nausea was gone.

Looking back on this later, it all seemed incredible. Forgive your father and you will feel better? It felt contrived, like the plot of some second-rate TV drama. He even tried to make himself believe it hadn't happened or, at any rate, that it was just the chance coincidence of bad ice cream and a tired body. Yet it was true. It had happened. However unsophisticated, however artless, however childish. It had happened.

Michael stood up, looked around. He removed his glasses, rubbed his eyes, replaced the glasses, and exhaled loudly. He went back upstairs, to his bedroom, changed, and climbed into bed. He opened Dante again, turning to the thirty-fourth canto: the descent to the bottom of hell, just prior to the ascent out of hell:

The Center

Satan

Exit

Dante and Virgil go to the very center of Hell, below the burning regions, coming finally to the frozen lake of Cocytus – the bottom pit of Hell. There, trapped eternally in the ice, is Satan. An immense Satan, a gigantic Satan. The road out of Hell requires them to hold onto Satan's enormous body, at the center of the Earth, as they continue their descent. (Even the ancient Greeks knew that the world was round.)

At first, Dante and Virgil continue climbing downward – literally down Satan's body – to the center of the Earth. Inexplicably, climbing *down* Satan's haunches transforms itself into climbing *up* Satan's legs, which now point upward. They are beginning their climb out.

And thus ends the Inferno:

> Down there, beginning at the furthest bound of Beelzebub's dim tomb, there is a place which is known, not by sight, but by the sound of a stream that descends there through the hollow of a rock which it has worn in its winding course and gentle slope. Virgil and I entered on that hidden path to return into the bright world, and without caring to have any rest we climbed up, he first and I second, so far that I saw through a round opening some of the fair things that Heaven bears; and thence we came forth to see again the stars.[9]

THE MORNING OF THE THIRD DAY

Sunday Morning

April 5, 1992

"Damn!" Michael jumped back out of the shower. Too cold! The shower was set up in the basement, attached to the central pillar. The furnace was a few feet away, as were the two tables. He tried to adjust the spigots. Not easy to do. Michael figured those spigots hadn't been replaced for decades. They were stiff as hell, but after a few more tries, the temperature was acceptable. Ever since reaching puberty, taking a shower down here had been a little bit strange. Here you were, right out in the middle of the basement. Naked as a jay bird. And even in the middle of summer, the concrete floor was cold.

Finally, after lathering up and rinsing down, Michael reached back to the spigots and shoved them both closed. Then banged them for good measure. A small drip finally stopped. He grabbed one of the towels from the clothes line and dried himself. Whew! *That* was a good way to wake yourself up in the morning. After dressing, he advanced his watch by one hour. Today was the beginning of Daylight Savings; it was now a little past 8:30 a.m. – Central Daylight Time. Since his body was still essentially on Eastern Standard

Time, it more or less felt like 8:30 to him; he was now in sync with the Midwest. Of course, when he got back home to Maryland, he'd have to adjust like everyone else. But he could put that off till tomorrow.

He walked over to the lounge chair at the base of the stairs and sat down. The furnace, scrubbed by the sun, was now just a dusty, fragile set of duct-taped tubes. Not even remotely ominous; unlike last night. The letter from Elgin. The suicide. The reprieve. The nausea. The forgiveness. Unreal, as if he'd watched someone else go through it. Or read about it. Or overheard a story about it at the Village Inn.

What to do now? Write a letter, maybe, to that woman in Elgin. Get her phone number. Talk to her. Scope out the situation and then figure out the next step. He'd draft the letter on the plane, on the return trip to Maryland.

Adult duty discharged for the moment, he thought about the last time he'd seen Dad. What was last night all about? The final healing of a very old psychological wound? Before this weekend, such an analysis would have pretty much exhausted the range of possible explanations. But now, there was the possibility that last night's episode also entailed something beyond the reach of the explicable. To be honest, it wasn't just a possibility, it was more like a certainty.

Explicable or not, though, it was time to let this wound heal on its own. No more about it today.

So, then, what *about* today, here, now? After sleeping on yesterday, two things stood out. First, there is a significant gap between the reach of reason and the scope of reality, attested by centuries of painstaking work by patient scientists. Work that unearthed the immense, ancient universe, and our token tenure in that universe. We are truly infants of infinity. This was the first and the clearest standout from yesterday. The second was his persistent yearning, the object of which – the 'something' – was now freed from the burden of heavy-handed clarity.

Again the question: What to do now? There was no urgency to clarify the first point, since it seemed pretty clear as it stood; blindingly obvious, to put it less cautiously. The work on the clarification of that yearning still lay ahead. Not only was there the work of clarification, there was also the work of how to do that work – how to clarify a 'something' that, by its very nature, would always resist clarification. A tough balancing act.

Before taking up that second question, Michael finally decided to face a suppressed feeling that had dogged him all weekend. Every once in a while, he'd sense a distinctly religious trend to his thought: A reality beyond reason? A reality that would always resist clarification? A reality seen *through a glass darkly*? A knowledge of the inexplicable that one might call, perhaps, *faith*? He finally said aloud to himself, "OK, Mr. Ex Methodist, don't you think you should resurrect your Christianity, at least long enough to seriously reconsider what's there?" The answer just had to be yes. Perhaps obvious. Perhaps, um, *blindingly* obvious.

Michael looked at his watch. Not quite nine-thirty. He knew from his mother's funeral, less than two years ago, that Sunday worship was at ten o'clock. He had plenty of time. As he rose from the chair and turned to climb the basement stairs, an idea hit him: maybe Schröder's staircase could serve as a metaphor of the relationship between things-explicable and things-inexplicable.

There is an old optical paradox known as Schröder's staircase: Is wall A closer to you, or is wall B?

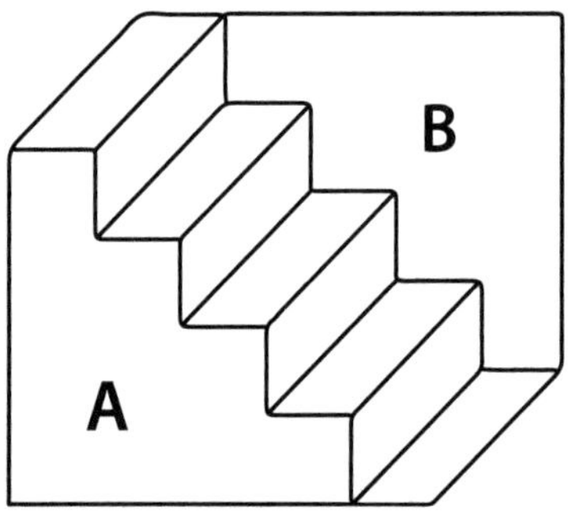

The ordinary world was the world in which you used stairs to go up and down, the world in which A was in front of B. If, for some reason, you ducked beneath the stairs – to build a spaceship, say, or a haunted house – you were in the B-in-front-of-A world, the world in which the shape of the stairs no longer served any ordinary, practical purpose. Perhaps the skill of shifting perspectives would become a standard skill (an ordinary skill, dare one say?) in a mature epistemology which openly and explicitly acknowledged the gap. When you spoke to someone, you were really addressing the B-in-front-of-A self, even though all you could see was the A-in-front-of-B body.

Michael continued up the A-in-front-of-B stairs.

⚜

A little before ten, Michael sat again in a back pew of Main Street Methodist Church. He had been here as a young pre-teen, initiated into formal membership on Palm Sunday, 1960. He had spent hours at home in his bedroom, transcribing scripture into Old English script with his new Speedball Calligraphy set. Sam the cat on his lap. Handel's *Messiah* playing on the record player in the background. Again. And again. And again. His appetite for repetition greatly exceeded his Grandfather's, who suggested that young Michael close the bedroom door, then went to the basement to work on something. Anything that would make noise. For a while, the house was in quite good repair, compliments of George Frederick Handel. Michael's discovery of Ravel's *Bolero* – a maniacally repetitive symphony – almost brought his poor Grandfather to the point of despair. He longed for a quiet Ohio farm.

Michael had become an altar boy at the Church, dutifully walking down the aisle every Sunday to light the candles, sit through the service, then extinguish the candles and walk back out, Reverend Henderson behind him. Michael became vice president of MYF, the Methodist Youth Fellowship. On the evening of Palm Sunday, 1962, just before an MYF meeting, he was talking to a friend at the north door. Noticing suddenly that it was after six – the meeting was to begin promptly at six – he raced through the dark sanctuary, taking a short-cut to the meeting room. He tripped going by the altar. There was a hard, sharp pain above his right eye. He got up and continued to run – noticing that he suddenly seemed to be sweating quite a lot. As he entered the room, all conversation abruptly stopped. Everyone was staring at him. He looked down. His white shirt had a large, bright red stain in the middle, and the stain was growing. He had neatly opened a gash in his forehead that was energetically pumping out blood; the large black eye that ensued would earn him the name of Altar Boy in his eighth-grade classes. A call to the Moonlight restaurant brought the family back to take him to the hospital. The Moonlight was a weekly ritual, Grandma's night off. They had just ordered dinner when the call came. Grandpa was exasperated. Now, decades later, Michael sat quietly at the back of church, looking at that same altar.

Sunday, November 4, 1990

IT WAS THE morning before his mother's viewing at Eunice Smith funeral home. Michael was in the Main Street Methodist church, talking with the assistant minister; the morning church service would begin in a few minutes. The assistant minister was trying to console him over the loss of his mother, two days earlier. "Michael," he had said, "even as we stand here, Virginia's spirit is looking down on us." Suddenly, Michael experienced an overwhelming vision – a flashback of himself and Carolyn seen through his mother's eyes. Or rather through the eyes of her departing spirit. It was a vision of Michael and Carolyn, seen from above, and moving rapidly toward them as they walked down the hospital corridor in the opposite direction, passing the Halloween decorations, moments before they were told of Code Blue and shown Virginia's lifeless body. The vision was from near the ceiling, moving toward Michael and Carolyn much more rapidly than they were approaching.

The vision almost undid Michael. A few minutes later, sitting in the back pew with Carolyn, he dug his fingernails into the palms of his hands to keep from breaking down. He deliberately tried to control his breathing.

A year and a half later, Michael again sat at the back of the church, waiting for the service to begin. He leafed through the program. On the back page was a prayer.

I asked God for strength, that I might achieve . . .
I was made weak that I might learn humbly to obey.
I asked for health, that I might do great things . . .
I was given infirmity, that I might do better things.
I asked for riches, that I might be happy . . .
I was given poverty, that I might be wise.
I asked for power, that I might have the praise of men . . .
I was given weakness, that I might feel the need of God.
I asked for all things, that I might enjoy life . . .
I was given life, that I might enjoy all things.
I got nothing that I asked for . . .
But everything that I had hoped for.
Despite myself, my unspoken prayers were answered . . .
I am among all men most richly blessed.

Most richly blessed. Someone who wanted health, riches and power; someone who wound up with infirmity, poverty, and weakness, yet counted it a blessing. Beneath the last verse was a note that this prayer had been found in a bible on a Civil War battlefield, handwritten on the inside cover. Some soldier – from the North or the South – may have read that very prayer the morning of the battle whose end he would not live to see. Now, today, Michael sat here in the pew of a southern Illinois church reading that very prayer.

Michael was musing over this when the worship service began. Brought back to the here and now, he looked at the program, trying to remember what he was supposed to do. There was a responsive reading. The minister would read the lowercase words, the congregation the uppercase:

CALL TO WORSHIP

L: The God of heaven and earth calls us together.

C: WE COME TO EXPLORE MATTERS OF THE SPIRIT.

L: There is so much we do not know or understand.

C: SO MUCH OF GOD IS UNKNOWN TO US.

L: Let us worship the One who gives us breath.

C: WE REACH FOR THE ONE BEYOND OUR KNOWING.

L: In God we live and move and have our being.

C: ALL THAT WE HAVE AND ARE COMES FROM GOD.

*HYMN OF PRAISE NO.278 "Open My Eyes, That I May See"

Everyone else stood up.

Oh yea, the "*". Stand when you see the "*". Michael stood and sang "Open My Eyes, That I May See." He had sung in junior high school chorus, and could still more or less sight read music. His thoughts began to drift. To this and that. Suddenly the people around him chanted, "And also with you." Michael snapped back to the present, turning the announcements over.

THE GREAT THANKSGIVING

L: The Lord be with you.

C: AND ALSO WITH YOU.

Nicely symmetric. The leader and the congregation were treated equally. As Michael resumed following the liturgy, he remembered another time, back when he really was an altar boy, sitting in the front pew while Reverend Henderson droned on and on with a sermon Michael couldn't understand or follow. It had been a warm summer day, and he kept thinking about everything he wanted to do outside. The service wound its way down. Michael mechanically stood and sat and stood and sang and chanted and sat. But his attention was not on what he was doing. Finally, there was a long silence. He glanced back to the row of people in the second pew on the other side of the aisle, just behind him. They were looking back at him, expectantly. Michael looked instantly to Revered Henderson, who was also looking at him, silent and smiling. *Damn it! The candle!* Michael shot up as if fired from a cannon, barely avoiding another fall. He put out the candle. And left it a good millimeter or two shorter and fatter.

Today, Michael was paying better attention. At one point, Revered Breece picked up the bread: "This is the body of Christ broken for you." Then the wine: "This is the blood of Christ, spilled for you." Then came communion.

GIVING THE BREAD AND THE CUP

** Please come to receive the sacrament by the center aisle.*
Return to your seat by the side aisle.

Michael dutifully joined the line forming in the center aisle leading to the altar. Michael smiled to himself: it was remotely possible that there were still a few molecules of *his* blood on that altar.

He returned to his seat and sat down. Reverend Breece stood. "Next Sunday, as you all know, is Palm Sunday. We will begin our observation of the traditional Holy Week with palms for everyone. Today, in preparation for our celebration, I want to go yet further back. All the way back. Back to some of the most basic scriptures of the Christian life. Back to the beginning. The first five verses of Genesis:

> ***In the beginning*** God created the heavens and the earth. Now the earth was formless and empty, darkness was over the surface of the deep, and the Spirit of God was hovering over the waters. And God said, "Let there be light," and there was light. God saw that the light was good, and he separated the light from the darkness. God called the light "day," and the darkness he called "night." And there was evening, and there was morning – the first day.

And the first five versus of John:

> ***In the beginning*** was the Word, and the Word was with God, and the Word was God. He was with God in the beginning. Through him all things were made, and without him not one thing was made that has been made. In him was life, and that life was the light for humankind. And the light shines in the darkness, and the darkness has not mastered it.

Breece proceeded to develop these opening lines, mostly with material from the Epistles of Paul. Not a favorite among Michael's friends back east, nor really a favorite of Michael's. Paul's doctrines flew directly in the face of much modern thought. Breece's explications left Michael unconvinced. Not necessarily opposed, just feeling that real difficulties were being discounted.

...except that the real problem wasn't logical quibbles with Paul. It was the small, still voice deep inside, whispering in disbelief, "Here I am in a church. In a church pew. In a church pew singing 'Open My Eyes, That I May See.' Do I really believe *any* of this?"

Michael left shortly after the service. He walked slowly down the front stair, wrapped in thought. The morning was cold, clear, and very windy, but not uncomfortable; last night's rain had moved on. He shoved his hands deep into his coat pockets and walked up Main Street, frowning. The sky was a bright, cloudless blue. The trees swayed in the strong wind. Wind gusts swept across the lawns, leaving waves of movement in their wake, as if invisible hands were lightly stroking the ground. He paused at the corner of Main and Edwards. Behind him was the Elias Smith Funeral Home. A short distance east along Edwards was Horace Mann, just beyond the intersection with Clawson.

He stood for a time, looking at the school building. He tried to recall the school year of 1955 – 1956. A brand spanking new Horace Mann. Several vague images came and went. Then his eyes widened a bit, as he remembered a little Halloween parade Mrs. Shea had organized in 1955 for his second grade class. Everybody brought their Halloween costumes to school, put them on, and marched out to parade around the block just west of the school, with Mrs. Shea and an assistant to help ride herd. Out the door, up Clawson to College, along College to Main. Then, back down Main to Edwards, and back to school. Though most of the memory was gone, two elements survived. First, the feel and the sound of those costumes. The crinkling, swishing coverall costumes, and the masks held on by a single white elastic strap. Those masks were awkward. The eyeholes were inadequate, and the trapped breath turned quickly stuffy. But the first memory was overshadowed by the second – a memory of sheer joy! All around Michael were animated witches, wolf-men, and skeletons. Even a bright orange pumpkin head, if memory served. Real, live spooks running around in front of Michael. The fantastical embodied and made real.

A seven-year-old's early taste of a later yearning for a world animated by things beyond the reach of normal human reason.

But stirring up memories was yesterday's work. Today, back to the real live world. Michael looked back up Main. Just across the street was Schwegel's Grocery. It'd been the A&P, before A&P folded. But more to the point, it appeared to be open. Michael decided to make breakfast at the house.

Strolling through Schwegel's, he decided on eggs, bread, butter, sausage, hash browns, and a new jar of instant coffee. There was grapefruit juice in the fridge back at the house; not his favorite, but it would do. Of course, he had to purchase more than he could eat. He'd just have to leave the remainder in the refrigerator, along with a note, indicating that this food had been purchased on the fifth of April. The new owner could judge for himself whether the eggs were still usable.

As he walked in the front door of the house ten minutes later, he knew what he'd say to Carolyn as soon as he got back: "Hi, dear. Guess what? Three whole days without a single vegetable! It was great!" He walked through the house to the kitchen, grinning impishly.

Before long, the radio was on KFUO, "Classical 99." Samuel Barber's Adagio for Strings was playing. Two eggs sizzled and spat butter. The water boiled and began to whistle and, at nearly the same instant, KACHUNK! Up popped a single slice of toast. Michael figured he could let that go and turned his attention to the boiling water, hastily making up a small pot of coffee. Turning back quickly to the stove, he picked up the skillet of eggs. Hmmm – not too bad. Even a little bit runny. Just right. He went back to the cupboard, pulled out a plain glass plate, dumped the toast onto the plate, and slid the eggs onto the toast. He wondered about salt, then remembered the hash browns and the sausage. They were no longer browning in the other skillet. They were burning.

Oh, well. They weren't *that* bad. He scraped the crispy sausage and the hash very-browns on top of everything, found the salt and pepper, and sat down. It all tasted absolutely wonderful.

As he ate, listening to the radio, he realized in a quiet way that his enjoyment of such music was a direct inheritance from his father. Guy would have enjoyed this music as much as Michael. Perhaps Guy *was* still enjoying the music. Perhaps Michael was, in a way, a reborn piece of Guy. The broken record had been resurrected.

Before long, finished with breakfast, Michael sat satisfied at the kitchen table, sipping his second cup of coffee. He sat in the chair in which he had always sat. Starting from his right, it had always been Mom, Grandma and Grandpa. Next week? Somebody else would be sitting here. Michael hoped he liked eggs. And sooner than later. He thought about leaving a little message taped on the back of his chair, like that on the toaster indicating which slot to use if there was only one piece of toast:

"When you're the only one in the house, sit here."

He leaned back in his chair and looked vacantly at the table in front of him. A bit of toast remained. As well, a bit of grapefruit juice. He ate the toast and drank the juice.

Michael had ceased to see the things around him. He stared blankly at the window in the front room, directly ahead of him. Then stood up and walked through the dining room, through the front room, and up the stairs – ducking his head at the last moment to avoid the low ceiling. He was just *never* going to get used to that low ceiling. Upstairs in the front bedroom, he walked to the window. The sun, though colder than yesterday, still poured in. The orange paint on the corner of the drawing table still glowed warmly. The sun filled everything. The sound of the wind shaking the house echoed in the empty rooms. The windows rattled. Old boards sighed. The stiff northwest winds were probably gusting to thirty miles an hour. It was like a storm of light. The sunlit outlines of the windows on the varnished wood floor and the painted walls wavered slightly, shimmering, as the windows moved slightly, enough to refract sun in pulsing, fluid patterns. Liquid sunlight.

He leaned against the window and looked across Washington Avenue through the old, rippled glass. The houses on the far side of the street seemed to sway slightly as the glass responded to the buffeting wind. Small bits of paper, moving slowly and erratically along the street, would suddenly spin up vertically several feet – momentarily entrained by a vortex of air – then fall back to their slow, erratic wanderings. He remembered something in Dante about the Sibyl – the prophetess of the ancient world – and the Sibyl's oracle who wrote prophecies on leaves, one letter to a leaf, then sent the message scattering on the wind. Presumably, the truth was contained in that strew, could one only gather the countless leaves and put the letters back into the right order. Returning to Alton, his life had felt like something that once made sense and then ceased to. But that could again. And perhaps after this weekend it might.

Michael picked up *Dante*, lying on the drawing board, and turned to *Paradiso*, the last of the three books of The Divine Comedy. He turned to the very end of its very last chapter. Dante was now in the highest level of heaven, looking directly at God:

> . . . for my sight, becoming pure, was entering more and more through the beam of the lofty light which in itself is true. From that moment my vision was greater than our speech, which fails at such a sight, and memory too fails at such excess. Like him that sees in a dream and after the dream the passion wrought by it remains and the rest returns not to mind, such am I; for my vision almost wholly fades, and still there drops within my heart the sweetness that was born of it. Thus the snow loses the imprint in the sun; thus in the wind on the light leaves the Sibyl's oracle was lost.[10]

Michael looked at his watch. About a quarter to one. His flight left Lambert St. Louis at 4:30. He looked back out the window for a long time. Then said, in a whisper, "Time to go." He turned and went back to his room, packed his clothes, made the bed and took the suitcase downstairs. He went back in the kitchen and finished cleaning the dishes. Finally, there was no more busywork to prevent him from recognizing he was about to leave the house for the last time. He felt his stomach tighten.

Michael went down to the basement, walked over to the train table. He wanted to touch everything, he wanted to... He didn't know *what* he wanted to do.

He walked around for a while, then slowly worked back through each room one last time, checking windows and locks. He placed the house key on the kitchen table and walked back through the dining room, through the living room, and out the front door. Suitcase in the back seat, he slid behind the wheel and headed back. Back down Washington Avenue. Back down Broadway. And in only ten minutes, he drove across the old Clark Bridge of the Mississippi river. A quarter mile downstream, its replacement was already under construction.

The Clark Bridge was the last bridge over the Mississippi before its confluence several miles downstream with the Missouri – though the technically correct word 'confluence' was somehow too smooth, too seamless to capture the reality of that crashing, heaving collision. The navigation channel in the river immediately upstream was posted with warnings to cling to the Illinois bank. At the confluence, the violent, swirling currents were so severe that all shipping was routed to a set of canals along the Illinois shore. This section of the Mississippi, one of the most turbulent of its entire course, was called the Chain of Rocks. The river would slow south of St. Louis. One hundred and fifty miles further south, at the southern tip of Illinois, it would join with the Ohio, changing from the 'upper' Mississippi to the 'lower' Mississippi. Wider and deeper. Never again as openly violent as the Chain of Rocks. Seven hundred miles further, past New Orleans, it would empty into the Gulf of Mexico. Some of its water would round the peninsula of Florida, mix with the Gulf Stream, and head northeast as one of the Earth's major ocean currents. Some of its water would remain in the Gulf. Some of its water would evaporate into Gulf winds. And some of that water in some of those winds would reach north, all the way back up to the upper Mississippi, there to fall back to earth as rain.

Rain. In a little over a year, the great flood of 1993 would engulf nearly everything. Nationwide news magazines would feature a satellite photo of St. Louis in which this slender peninsula separating the two rivers was gone – replaced by a dark, ominously wide ribbon of water. Beginning in June of 1993, the river would rise slowly to nearly fifty feet above flood stage. Most natural disasters are sudden and loud; a flood is neither. It would start as simple rain.

Over many weeks, the river would rise slowly but inexorably, until even the threat of more rain would raise everybody's anxiety. The river would breach people's dreams.

They called it a five-hundred-year flood. Something you expect once in five hundred years. The builders of the new Clark Bridge had planned to finish in the summer of 1993. Then came the flood. Ironically, the old Clark Bridge was one of the last closed by the flood. And the bridge survived the flood intact, as if to show itself dutiful to the last.

Given a reprieve by a five-hundred-year flood, the old bridge saw yet another autumn. And a winter. And a spring. But finally, the new bridge was completed. Then, one sunny day in the summer of 1994, unheralded, observed by only a few startled passers-by (the contractor having deliberately kept the exact time of demolition secret to avoid bothersome crowds) the old Clark Bridge came to an end. Puffs of smoke enveloped the main span. Slowly – almost in slow motion – the roadway and the old, rusted arch began to fall. A wave surged out as the severed span hit the water. For a minute, an image of the bridge remained in mid-air, traced out by the smoke of the explosions. Then the smoke cleared. The ghost of the bridge was gone. The water covered all but the angular upper section of the gray steel arch.

Epilogue

Deep space and deep time can provide more than just the springboards for the yearnings of a fledgling aerospace engineer. They can provide the principal premises in an argument for the inherent limitation of human reason. The basic argument is quite simple. Knowledge requires both time and ability. Weighed in the spatial and temporal scales given us by deep space and deep time, our planet-bound experience combined with our brief tenure imply that our knowledge of reality is tiny indeed compared with what there is to know. Thus, there is a significant gap between the reach of our knowledge and the scope of reality. And as reason is the principal faculty by which we acquire knowledge, one might more aptly say there is a significant gap between the reach of reason and the scope of reality.

But does the reality beyond the reach of reason have an impact on us or is it inconsequential? Do we see through a glass darkly, or not at all? If the former, then there should be *some* evidence of inexplicable events, presumably grounded in such beyond-reason reality. Is there such evidence?

I think there is, but it presents us with an interesting challenge: seeing something inexplicable *as* inexplicable, not as just another brute fact of existence. It requires us to see something old in a new way. It's evidence hiding in plain sight.

A plausible example of such evidence: the tension between the objective and the subjective, exemplified by the self. The key is to see this tension *as* inexplicable, not as a brute fact, nor as a temporary tension which will yield to some scientific explanation in the future – in other words, something only temporarily inexplicable. An even better example (for me, at least): my yearnings for a vague 'something', originally evident as yearnings for deep space and deep time. Uncovering this second example was quite a surprise. To explain it requires a digression.

When it became clear that I had to reexamine my negative opinions on religion, I turned to C. S. Lewis, arguably the preeminent Christian apologist of the twentieth century. I first came across his science fiction works long ago; at the time, I suspect I'd just have been puzzled by him. Much more recently, I began to read Lewis's apologetical works. I was encouraged when I came across, "I am not asking anyone to accept Christianity if his best reasoning tells him that the weight of the evidence is against it".[11] Even more encouraging:

> That which I greatly feared had at last come upon me. In the Trinity Term of 1929 I gave in, and

> admitted that God was God, and knelt and prayed: perhaps, that night, the most dejected and reluctant convert in all England.[12]

Here was a guy I could trust. Then, about five years ago I was reading the opening chapters of *The Pilgrim's Regress* – the first book Lewis wrote after his acceptance of Christianity. He gave an account of a feeling very much like my yearning for a vague 'something.' In the second chapter, John (Lewis's stand-in) saw a vision in a mysterious window in a wood. The description of the vision was not what grabbed me, it was his description of how it made him feel:

> There was no glass in the window and no bars; it was just a square hole in the wall. Through it he saw a green wood full of primroses: and he remembered suddenly how he had gone into another wood to gather primroses, as a child, very long ago – so long that even in the moment of remembering the memory seemed still out of reach. While he strained to grasp it, there came to him from beyond the wood a sweetness and a pang so piercing that he instantly forgot his father's house.[13]

It was at this moment I knew that, finally, I was hot on the trail of the answers to the questions I posed in my preface. (I was listening to a tape of *Pilgrim's Regress* during a long drive to New Jersey. I immediately started laughing and hitting the roof of my car. Fortunately, I didn't hit anything else.) Lewis's account of his desire – what I call my yearning – is an important part of his apologetics. For me, the key insight was the exceptional nature of the *something* desired. For Lewis, it wasn't just mysterious; it was never really seen *at all.* It was always just out of sight, just over the hill.

Two accounts of Lewis's desire stand out to me. In both, the ostensible identity of the 'something' was heaven but, *ex hypothesi,* it was never really *just* that, it was just *something.* In June of 1942 Lewis preached a sermon in the Church of St. Mary the Virgin, in Oxford. It was only six months after Pearl Harbor; and still far from clear that Hitler would not eventually overwhelm England.

> In speaking of this desire for our own far-off country, which we find in ourselves even now, I feel a certain shyness. I am almost committing an indecency. I am trying to rip open the inconsolable secret in each one of you – the secret which hurts so much that you take your revenge on it by calling it names like Nostalgia and Romanticism and Adolescence; the secret also which pierces with such sweetness that when, in very intimate conversation, the mention of it becomes imminent, we grow awkward and affect to laugh at ourselves; the secret we cannot hide and cannot tell, though we desire to do both. We cannot tell it because it is a desire for something that has never actually appeared in our experience. We cannot hide it because our experience is constantly suggesting it, and we betray ourselves like lovers at the mention of a name. Our commonest expedient is to call it beauty and behave as if that had settled the matter. Wordsworth's expedient was to identify

> it with certain moments in his own past. But all this is a cheat. If Wordsworth had gone back to those moments in the past, he would not have found the thing itself, but only the reminder of it; what he remembered would turn out to be itself a remembering.
>
> These things – the beauty, the memory of our own past – are good images of what we really desire; but if they are mistaken for the thing itself they turn into dumb idols, breaking the hearts of their worshippers. For they are not the thing itself; they are only the scent of a flower we have not found, the echo of a tune we have not heard, news from a country we have never yet visited.[14]

His sermon's reference to Nostalgia and his critique of Wordsworth's 'cheat' especially drew my attention. Another account of this desire – his best, I think – is given in the last chapter of *The Problem of Pain*:

> There have been times when I think we do not desire heaven but more often I find myself wondering whether, in our heart of hearts, we have ever desired anything else. You may have noticed that the books you really love are bound together by a secret thread. You know very well what is the common quality that makes you love them, though you cannot put it into words: but most of your friends do not see it at all, and often wonder why, liking this, you should

also like that. Again, you have stood before some landscape which seems to embody what you've been looking for all your life; and then turned to the friends at your side who appear to be seeing what you saw – but at the first words a gulf yawns between you, and you realize that this landscape means something totally different to them, that they are each pursuing alien visions and care nothing for the ineffable suggestion by which you were transported. Even in your hobbies, has there not always been some secret attraction which the others are curiously ignorant of – something, not to be identified with, but always on the verge of breaking through, the smell of cut wood in the workshop or the clap-clap of water against the boat's side? Are not all lifelong friendships born at the moment when at last you meet another human being who has some inkling (but faint and uncertain even in the best) of that something which you were born desiring, and which beneath the flux of other desires and in all the momentary silences between the louder passions, night and day, year by year, from childhood to old age, you are looking for, watching for, listening for? You have never had it. All things that have ever deeply possessed your soul have been but hints of it – tantalizing glimpses, promises never quite fulfilled, echoes that died away just as they caught your ear. But if it should become really manifest – if there ever came an echo that did not die away but swelled into the sound itself – you would know it. Beyond

> all possibility of doubt you would say 'Here at last is the thing I was made for.' We cannot tell each other about it. It is the secret signature of each soul, the incommunicable and unappeasable want, the thing we desired before we met our husbands or wives, made our friends or chose our work, and which we shall still desire on our deathbeds, when the mind no longer knows husband or wife, friend or work. While we are, this is. If we lose this, we lose all.[15]

Lewis plainly asserts that this desire is a desire for something we have never had, because that 'something' was itself never really visible: "You have never had it. All things that have ever deeply possessed your soul have been but hints of it – tantalizing glimpses, promises never quite fulfilled, echoes that died away just as they caught your ear." I submit that this is a breathtakingly apt description of a reality beyond the gap.

Once alerted, I began to identify other instances in which I had experienced something similar. As an adolescent, for example, I remember feeling an excitement at the beginning of many science fiction novels: I somehow expected to uncover something mysterious, something I could never quite put into words. It never happened. Even though there were novels I really liked, I clearly remember a sense of disappointment when I failed to find that vague 'something' I was looking for. It wasn't an overpowering disappointment; it just popped up enough that I knew it to be the renewal of an earlier feeling. In other words, it was a disappointment very much like Lewis's desire. In a very different sphere, adagios have always touched me in a way I now recognize as similar to Lewis's desire.

This unfocusable yearning, this desire, has served as the grounds for an old argument for the existence of God called (appropriately enough) The Argument from Desire[16] – an argument dating back to mediaeval times. Lewis put his version quite succinctly:

> Creatures are not born with desires unless satisfaction for those desires exists. A baby feels hunger: well, there is such a thing as food. A duckling wants to swim: well, there is such a thing as water. We feel sexual desire: well, there is such a thing as sex. If I find in myself a desire which no experience in this world can satisfy, the most probable explanation is that I was made for another world.[17]

At first blush, this argument seemed to me quite unconvincing. In the same vein, one could argue that the actual existence of Superman is "the most probable explanation" for the desire of adolescent boys to be Supermen. This is an ordinary desire for a (more or less) ordinary object. As such, ordinary cultural and psychological analyses are quite adequate for its explanation. These analyses employ normal reason within its normal scope. And these analyses unequivocally rule out the possibility that there really is a superman.

Even if I were to give the phrase 'ordinary objects' quite an extraordinary scope, a scope which would include, say, subatomic objects – a level at which even the word 'object' is problematic – an appropriately modified argument from desire still wouldn't work. We tolerate the bizarre, outlandish pronouncements of quantum mechanics *only* because they consistently lead to empirical predictions which have been verified again and again by empirical measurements. In other words, we tolerate them because they can finally pass muster in the everyday ordinary world, because they are within the reach of ordinary reason – a faculty quite capable of growth and development. Thus, even under this very flexible definition of ordinary, the argument from (ordinary) desire would fail.

I think Lewis erred in not making it plain that the desires of which he spoke were "tantalizing glimpses of promises never quite fulfilled, echoes that died away just as they caught your ear." Desires for 'something' that always resisted clarity. The Argument from Desire appears weak because we naturally apply it to ordinary desires for ordinary objects. I therefore submit that the Argument from Desire can pass muster when we realize that the 'objects' of desire are beyond the gap, objects whose reality is beyond the reach of (ordinary) reason.

So the best evidence for reality beyond the reach of reason – reality beyond the gap – turned out to be the very yearnings whose objects I had long been struggling to clarify, to get into focus! Yearnings first aroused in me by deep space and deep time.

I will state here, for the record, that Lewis has convinced me to return to my original Christianity. I will also state for the record that I still cannot quite believe I just wrote the words I just wrote. But to go further would be to begin the prologue of another book, not to end the epilogue of this one.

I will end this epilogue with a summary. The vague feelings aroused in me by deep space and deep time led me to some very definite goals – college degrees in aerospace engineering, for instance. But there was always a nagging anxiety: though I knew these vague feelings were about 'something', it was a 'something' I could never really understand. (Not surprising, perhaps, for vague feelings.) From philosophy, however, I learned that if I were to say I knew something, my knowledge claim would have to pass some pretty strict standards. Fair enough; in general, I had no objection. But wandering the streets of Upper Alton after my mother's death, I found myself struggling to find words to express a mysterious significance – a mysterious *something* – that seemed to hide under every stone, to hang in the humid night air. Peeling back its deeply emotional skin, I knew the core 'something' was – somehow – a real something. Something that existed outside me.

The problem? This 'something' resisted clarification. It seemed – by its very nature – to be unclear and unfocused. Yet the philosopher in me (very reasonably) insisted I could only know this *something* to be real, to be outside myself, if I could make it clear and bring it into focus. In other words, only if my knowledge claim could pass muster.

Thus the problem. The solution turned on the discrepancy between the reach of reason (and knowledge) and the scope of reality – a discrepancy so drastic as to defy description in language. And unless reality was governed by the cosmic equivalent of "what you don't know can't hurt you", human experience was pretty much guaranteed to include things-inexplicable, yet nonetheless real – things forever unclear and unfocused. Things like a vague *something*. Things which, by definition, would never satisfy the strict criteria of ordinary knowledge – the knowledge of things-explicable.

Therefore, there was a real, philosophically defensible possibility that what I experienced walking the streets of Upper Alton was connected to something real. More concisely, that it *was* real. Most surprising (and gratifying) of all was how all of this dovetailed with an idea developed by C. S. Lewis from an ancient bit of medieval theology. Thus, the vague 'something' whose pursuit drove this entire book wound up looking a lot like what Lewis described seventy years ago.

I have only begun to think through how to bring together the knowledge of things-explicable and things-inexplicable under some sort of philosophical super umbrella. I strongly suspect that the knowledge of things-inexplicable is going to end up looking a lot like faith. Seeing through a glass darkly. If my ignorance of faith has led me to fuss over what is routinely covered in a typical undergraduate theology seminar, I can only beg forgiveness.

Acknowledgments

After twenty years, it is difficult to assemble an adequate list of those who have helped me. Two people from Alton spring immediately to mind: David Scott (aka Bernie) and Cecil Griesbaum (aka Roy). Both helped me in ways nobody else could – with this book, and in the earlier life it chronicles. In Columbia (my current home) Gerry Goethe, Jean Link, Marie Moore, and Ruth Smith read and commented on an early version; John Lobel was both an attentive reader and a close friend.

Two people deserve special mention: first, Kevin Dungey. Kevin and I have been friends since he and I were on the faculty at St. John's College. Though neither of us were formally trained in philosophy (Kevin studied mediaeval literature at Stanford) we both have a deep interest in philosophy; Kevin gave me detailed critiques of my philosophical arguments. He has consistently been there for me. When he and a business partner recently started a publishing company, I immediately knew whom I wanted to publish this book.

The second person has been, without doubt, my most consistent supporter: my wife, Bonnie Orrison (aka Carolyn). If I had not met and married Bonnie, this book would never have been written. I walked straight into unemployment twenty years ago to write it. For most of that time, I was able to kick in a little something for my room and board by teaching philosophy and, later, engineering science as an adjunct instructor at three local colleges – absolutely wonderful jobs, but not especially lucrative. I owe Bonnie more than I can ever repay, and I don't just mean financially.

David Irvine
Columbia, Maryland
November 2, 2016

Endnotes

1 Eiseley, Loren. 1957. "The Slit." *In The Immense Journey.* New York: Vintage Books. p 3.

2 Alighieri, Dante. 1939. *The Divine Comedy: Inferno.* Translated by John D. Sinclair. New York: Oxford University Press. p 23.

3 Virgil. 1953. *The Aeneid.* Translated by C. Day Lewis. New York: Doubleday Anchor Books. p 13.

4 Alighieri, Dante. 1939. *The Divine Comedy: Inferno.* Translated by John D. Sinclair. New York: Oxford University Press. p 37.

5 Norton, Andre. 1953. *Star Rangers.* New York: Harcourt Brace. p 3.

6 Descartes, René. 1984. *Meditations on First Philosophy. In The Philosophical Writings of Descartes, Volume II.* Translated by John Cottingham. New York: Cambridge University Press. p 12.

7 Descartes, René. 1984. *Meditations on First Philosophy. In The Philosophical Writings of Descartes, Volume II.* Translated by John Cottingham. New York: Cambridge University Press. p 17.

8 Pascal, Blaise. 1966. *Penseés.* Translated by A. J. Krailsheimer. London: Penguin. p 85.

9 Alighieri, Dante. 1939. *The Divine Comedy: Inferno.* Translated by John D. Sinclair. New York: Oxford University Press. p 427.

10 Alighieri, Dante. 1939. *The Divine Comedy: Paradiso.* Translated by John D. Sinclair. New York: Oxford University Press. p 481.

11 Lewis, Clive Staples. 1958. *Mere Christianity.* New York: Macmillan. p 108.

12 Lewis, Clive Staples. 1955. *The Surprised by Joy.* In: *The Inspirational Writings of C. S. Lewis.* 1991. Inspirational Press: New York. New York. p 125.

13 Lewis, Clive Staples. 1943. *The Pilgrim's Regress. In: The Collected Works of C. S. Lewis.* 1996. Inspirational Press: New York. p 12.

14 Lewis, Clive Staples. 1965. "The Weight of Glory." In: *The Weight of Glory and Other Addresses.* Grand Rapids, Michigan: William B. Eerdmans Publishing. p 4.

15 Lewis, Clive Staples. 1962. *The Problem of Pain.* New York: Macmillan. p 145.

16 Bassham, Gregory, ed. 2015. *C. S. Lewis's Christian Apologetics: Pro and Con.* pp 27-44.

17 Lewis, Clive Staples. 1958. *Mere Christianity.* New York: Macmillan. p 106.

COLOPHON

Printed on 50# Glatfelter Offset B18 Eggshell

Set in Adobe Garamond Pro

Design & Typesetting by Joshua Mitchell